Universal Command and Control Language Early System Engineering Study

Performance Effects of a Universal Command and Control Standard

JAMES DIMAROGONAS, JASMIN LÉVEILLÉ, JAN OSBURG,
SHANE TIERNEY, BONNIE L. TRIEZENBERG, GRAHAM ANDREWS,
BRYCE DOWNING, MUHARREM MANE, MONICA RICO

Prepared for Office of the Secretary of Defense
Approved for public release; distribution unlimited

NATIONAL DEFENSE RESEARCH INSTITUTE

For more information on this publication, visit **www.rand.org/t/RRA744-2**.

About RAND

The RAND Corporation is a research organization that develops solutions to public policy challenges to help make communities throughout the world safer and more secure, healthier and more prosperous. RAND is nonprofit, nonpartisan, and committed to the public interest. To learn more about RAND, visit www.rand.org.

Research Integrity

Our mission to help improve policy and decisionmaking through research and analysis is enabled through our core values of quality and objectivity and our unwavering commitment to the highest level of integrity and ethical behavior. To help ensure our research and analysis are rigorous, objective, and nonpartisan, we subject our research publications to a robust and exacting quality-assurance process; avoid both the appearance and reality of financial and other conflicts of interest through staff training, project screening, and a policy of mandatory disclosure; and pursue transparency in our research engagements through our commitment to the open publication of our research findings and recommendations, disclosure of the source of funding of published research, and policies to ensure intellectual independence. For more information, visit www.rand.org/about/principles.

RAND's publications do not necessarily reflect the opinions of its research clients and sponsors.

Published by the RAND Corporation, Santa Monica, Calif.
© 2023 RAND Corporation
RAND® is a registered trademark.

Library of Congress Cataloging-in-Publication Data is available for this publication.
ISBN: 978-1-9774-0866-2

About This Report

The U.S. Department of Defense (DoD) requires more efficient and timely methods to acquire, integrate, and interoperate systems and, perhaps more crucially, systems-*of-*systems (SoSs) to deter near-peer adversaries in a rapidly evolving threat environment and prevail in combat should deterrence fail. To meet this need, DoD has undertaken many initiatives to create command and control (C2) standards for interoperability of its own systems, U.S. civilian systems used in homeland defense, and allied systems used in coalition fights. In 2019, the RAND Corporation's National Defense Research Institute was asked to participate in a multiyear effort to help DoD understand the challenges of creating a universal C2 language to facilitate the evolution of systems and interoperability of SoSs. Striking the right balance among competing standardization objectives is challenging, and these design choices have the potential to adversely affect weapon system performance. In this report, we establish a conceptual framework for analyzing SoS performance of different sensor-to-shooter connections, combinations, and associated C2 constructs. The intent is not to accurately estimate the performance of a specific SoS with and without a universal interface, but rather to explore the range of trade-offs designers make between performance and characteristics of a standard interface, using examples from active protection systems, electronic warfare, and ballistic missile defense.

The research reported here was completed in December 2021 and underwent security review with the sponsor and the Defense Office of Prepublication and Security Review before public release.

RAND National Security Research Division

This research was sponsored by the Office of the Secretary Defense and conducted within the Acquisition and Technology Policy Center of the RAND National Security

Research Division (NSRD), which operates the National Defense Research Institute (NDRI), a federally funded research and development center (FFRDC) sponsored by the Office of the Secretary of Defense, the Joint Staff, the Unified Combatant Commands, the Navy, the Marine Corps, the defense agencies, and the defense intelligence enterprise.

For more information on the RAND Acquisition and Technology Policy Center, see www.rand.org/nsrd/atp or contact the director (contact information is provided on the webpage).

Contents

About This Report iii

Figures vii

Tables ix

Summary xi

Acknowledgments xix

Abbreviations xxi

CHAPTER ONE

Introduction 1

C2 Interface Standardization Challenges 1

Organization of This Report 3

CHAPTER TWO

Methodology 5

Approach 5

Assumptions, Constraints, and Limitations 9

CHAPTER THREE

Standards 11

Standards to Achieve Interoperability 11

DDS–eXtremely Resource Constrained Environment—An Example of the
  Trade-offs Required for Performance 19

Nontechnical Considerations for Standardization Efforts 25

Conclusions 28

CHAPTER FOUR

Insights from Testing and Experimentation 31

STITCHES Performance 32

OMS Performance 32
Data Distribution Service Performance 33
Analyzing Universal Command and Control Interface and Variable Message Format 36
Potential Inefficiencies of a UCCL... 37
Summary of Testing and Experiment Data Review 38

CHAPTER FIVE
Electronic Warfare Mission Thread.. 41
Modeling the Mission Thread and Its Interfaces..................................... 42
Modeling Results and Critical Step Analysis 46
Conclusions.. 49

CHAPTER SIX
Active Protection System Mission Thread 51
Modeling the Mission Thread and Its Interfaces.................................... 51
Modeling Results/Critical Step Analysis 59
Conclusions.. 64

CHAPTER SEVEN
Ballistic Missile Defense Mission Thread.. 67
Modeling the BMD Mission Thread... 70
Worst-Case Scenario for Impact of Latency on Medium-Range Ballistic
      Missile Tracking.. 78
Critical Step Analysis.. 81
The Mission Thread and Its Interfaces... 82
Conclusions... 91

CHAPTER EIGHT
Conclusions... 93

APPENDIX
Publish and Subscribe Overview 97

References.. 99

Figures

2.1. High-Level Mission Threads C2 Model 7
3.1. The OSI and Internet Models of Communications 13
3.2. Bit Structure of an Internet Header 15
3.3. Encapsulating for Interoperability Increases the Amount of Information
That Must Be Passed Between Nodes in a System 19
4.1. DDS-1 Scalability 35
5.1. Electronic Warfare Mission Thread High-Level Operational Concept 43
5.2. UCCL Integration Results: Maximum Jammable Distance 48
6.1. Cruise Missile Threat Scenario 53
6.2. RPG Scenario 53
6.3. MDD as a Function of Operations per Bit and Header Size for
Cruise Missile Threats 60
6.4. MDD Breakdown Across Phases of Engagement for Cruise Missile Threats 61
6.5. MDD as a Function of Operations per Bit and Header Size for
RPG Threats 62
6.6. MDD Breakdown Across Phases of Engagement for RPG Threats 63
7.1. The BMD Kill Chain 68
7.2. Mission Thread Delay Versus Computational Complexity Factors 75
7.3. Mission Thread Delay Versus CPU Speed 76
7.4. Single Step Delay, in Seconds, as a Function of Link Data Rate and
Interface Implementation 77
7.5. Flight Profile of an MRBM 79
7.6. Error in Target Position for Different Amounts of Latency 80
7.7. Communication Patterns for Threat Identification, Track Estimation,
and Downstream Sensor Cueing in Current BMD Architecture 83
7.8. Communication Patterns for Target Discrimination in Current BMD
Architecture 85
7.9. Communication Patterns for Interceptor Engagement in Current BMD
Architecture 87
7.10. Key Sensor Information Flows for a Future BMD Architecture 89

Tables

3.1. Protocol Design Trade-offs, According to the Situation in Which Each
       Returns Benefits 24
4.1. Results of DDS Architecture Investigations as Reported by Xiong et al.,
       2011: Latency and Jitter, by Implementation 34
4.2. Results of DDS Architecture Investigations as Reported by Xiong et al.,
       2011: Bandwidth, by Implementation and Communications Model 35
5.1. Modeling Parameters 46
6.1. Threat Types 58
6.2. Modeling Parameters 59
6.3. Four Data Formats 60
6.4. Definitions of the Phases of the MDD 62
7.1. Modeling Parameters 74
7.2. Mapping BMD Mission Functions to Physical Systems 82

Summary

Background

Command and control (C2) of military capabilities in the late 20th and early 21st centuries require linking, synchronizing, and directing multiple complex systems under tight time constraints, all while under attack by the enemy. The consequences of failure are severe. At the tactical and operational levels, systems that do not work well together cannot bring their full capabilities to bear or may even malfunction and cause collateral damage. At the strategic level, a malfunctioning system of systems (SoS) can result in catastrophic battlefield losses.

The U.S. Department of Defense (DoD) has therefore undertaken many initiatives to create and document C2 standards for interoperability.[1] Standardized, well-documented and well-understood interfaces ensure that the systems involved exchange the right data in the right format and interpret those data in the right way. However, standardizing, documenting, and educating personnel in the use of an interface is not enough to guarantee interoperability. Interfaces that are designed for a particular context are optimized for a mission and for the systems that support that mission. Assumptions about the mission—and the operational concepts used—are embedded within the interface optimization process. While this is adequate for missions that do not involve large numbers of different sensors and shooters and where the context is reasonably bounded and understood, the problem becomes intractable when one attempts to allow any sensor to connect to any shooter, each of which was designed to its own set of mission assumptions, constraints, and operational concepts. This leads to the typical n-squared interface problem in systems engineering that increases com-

---

[1] For the purposes of this report, we will use the International Organization for Standardization/International Electrotechnical Commission 25010 definition of *interoperability*: "Interoperability is the degree to which two or more systems, products or components can exchange information and use the information that has been exchanged."

plexity and cost in developing and maintaining interoperable interfaces and potentially reduces reliability. Creating a common interface standard for the SoS that is capable of expressing the nuances needed for interoperability while also meeting all the performance objectives of the mission is challenging. Striking the right balance between the competing standardization objectives of interoperability and performance is perhaps more art than science.

With the above in mind, the Office of the Under Secretary of Defense for Research and Engineering in 2019 asked the RAND Corporation's National Defense Research Institute (NDRI) to research key aspects related to the implementation of a still-to-be-designed Universal Command and Control Language (UCCL) and assess the effects that a UCCL might have on the performance of weapon systems and SoSs.

Research Objective

Our research objective is to establish a conceptual framework for analyzing SoS performance of different sensor-to-shooter connections, combinations, and associated C2 constructs. The intent is not to accurately estimate the performance of the system with and without a universal interface, but rather to explore the range of trade-offs by identifying how performance depends on the characteristics of interfaces and how it varies with respect to the details of the technical implementation. Consequently, this effort should not be viewed as a study of a specific standard interface but as an early system engineering study of how such an interface could and should be designed.

In this report, we review several DoD SoS integration efforts to understand their approach to achieving interoperability and the impact on performance (latency, throughput) of SoSs that use that approach. We also review a commonly used open interface standard, the Data Distribution Service (DDS), that strives to balance interoperability concerns with performance and describe the trades that were made in producing a version of the DDS that is lightweight enough to push out to the edge computing devices commonly known as the *Internet of Things* (IoT). With that background, we then identify the performance requirements for each of three DoD missions. Using a mission thread analysis, we assess how imposing a common UCCL might impact the performance of those missions.

What We Found

Insights from Ongoing DoD SoS Integration Efforts

The specific experiments we reviewed examined the following interface standards, all of which are used in DoD weapon systems: SoS Technology Integration Tool Chain for Heterogeneous Electronic Systems, Open Mission Systems, Universal C2 Interface, DDS, and Variable Message Format.[2] The following are the findings from this review:

1. Interface technical performance—as measured by throughput/bandwidth and latency of processing the data packets—is greatly affected by implementation details, such as message encoding and compression. It is not merely the definition of the interface but also the implementation that matters to performance. Different implementations of a standard can result in different packet sizes in real-life implementations, including fixed format binary, general purpose binary, compressed Extensible Markup Language (XML), and regular XML. We use this range of packet/data size as a proxy for different interface implementations in our modeling of specific mission threads.

2. Features of the network in which the standard is installed—such as architecture, communications model, and number of participants—can have significant impact on system performance, even when the standard and its implementation are unchanged.

3. Higher complexity associated with processing of some messaging standards has a significant effect. Central processing unit (CPU) processing time needed for packet processing (encoding, encryption, decryption, decoding, and processing of the information) is a significant contributor to overall performance of an interface.

Insights from Mission Thread Analysis

To gain insights into the effect of a UCCL on operational performance, we conducted analyses of three mission threads chosen to highlight a range of different military operations with varying sensitivity to message delays (i.e., latency):

1. An electronic warfare (EW) mission based on the CONverged Collaborative Elements for RF [Radio Frequency] Task Operations (CONCERTO) system

---

[2] Note that we did not perform independent experiments of these interface standards but confined our work to the review of current experiments conducted by reputable entities.

developed for the Defense Advanced Research Projects Agency (DARPA). This is a single system thread with extremely short timelines and very simple C2 exchanges between subsystems.

2.   An active protection system thread based on the Marine Air Defense Integrated System (MADIS), in which a few platforms coordinate a local defense against incoming threats with relatively short timelines.

3.   A ballistic missile defense thread. This thread has reasonably long mission execution times and a fairly complex C2 construct with multiple sensors, shooters, and C2 elements.

All three of these SoSs have already made a basic trade—the complexity of the C2 is reduced as the needed response times shorten. Yet by studying their implementation in more detail, we may learn exactly why and how mission performance is affected by the performance of the "language" of its interfaces. Our analysis shows that implementation details of a standard interface may contribute to interface overhead that changes technical performance by orders of magnitude. The most pronounced effect is due to delays in the interfaces. However, technical performance does not necessarily translate to impact in mission performance.

Electronic Warfare Mission Thread

CONCERTO is an EW system designed to minimize size, weight, and power when installed in military aircraft. EW requires very short reaction times, and the CONCERTO system is optimized for speed. The key technical measure for mission success is the maximum distance at which the aircraft can successfully jam a target emitter— this is warfare at the speed of light. We found that

- adding even a modest amount of interface overhead will affect mission performance, but in some cases, the effects could be mitigated.
- if the overhead increases above that modest amount, significant operational performance degradation is likely. These types of submillisecond system responses require a very carefully tailored interface or a much-reduced overhead version of a UCCL.
- implementing standard interfaces for systems used in EW is high-risk and requires careful and detailed engineering analysis to ensure that performance trade-offs do not adversely affect mission success.

Active Protection System Mission Thread

The MADIS is a family of systems providing air defense capabilities to protect a ground maneuver force on the move against such threats as unmanned aerial systems and fixed-wing and rotary-wing aircraft. MADIS is being developed as the first of three increments toward modernizing the Marine Corps' existing ground-based air defense capability. The key technical measure for mission success is the minimum distance at which a threat can be detected and still defeated, i.e., the *minimum defeat distance*. Our analyses indicated that

- *latency* (i.e., delay) caused by a UCCL can influence operational performance, but the impact may be modest
- in most cases, interface inefficiencies are not likely to be the main contributor to the minimum defeat distance
- for stressing cases of higher-end threats and tight performance requirements, careful engineering analysis should be performed to ensure operational performance trade-offs do not compromise mission success.

Ballistic Missile Defense Mission Thread

Missiles designed to fly a ballistic trajectory threaten territory far from their launch sites and are hard to defeat but have key periods of vulnerability that a layered defense system exploits. The ballistic missile defense (BMD) mission thread takes minutes to execute from launch detection to final kill. We found that

- additional end-to-end latencies on the order of seconds imposed by an inefficient interface design are unlikely to influence the successful execution of the thread
- however, some links within the thread have very constrained bandwidth—such as the downlink from a satellite detecting launch of the missile, or the fire control link to the interceptor—that might require an optimized UCCL interface[3]
- a UCCL may provide the opportunity to rapidly add sensors to the missile defense system, enabling a diverse set of sensor geometries that improve trajectory estimates and warhead discrimination, improving P_{kill}. However, throughput and

---

[3] For example, we identified the final sensor update to the interceptor as a critical step in the thread that could impact probability of kill (Pkill). Latency in this step caused by a UCCL is *unlikely* to influence Pkill if elements of the system have been calibrated for clock drift and if interfaces carry critical timing information for each element of the kill chain. We caution, however, that it is probable that latency will influence warhead discrimination. This unclassified study did not model warhead discrimination, but these algorithms are known to be latency-sensitive.

congestion of the underlying network by adding more sensors using an inefficient UCCL could erase these gains.

- if real-time composability of the sensor-to-shooter network is desired, network throughput and congestion are *likely* to be greatly affected by a UCCL's discovery protocol. The detailed implementation of a discovery protocol should be tailored to mission needs to account for bandwidth-constrained elements of the system.

Conclusions and Implications

In conclusion, our research demonstrates that implementation details of interfaces may change technical performance by orders of magnitude; the most pronounced effects we found are caused by delays in the exchange of information. Therefore, designers of interfaces may need to trade off technical performance to achieve interoperability.[4] However, we also find that technical performance does not necessarily translate to an impact on mission performance.

From a mission performance perspective, we found that

- mission performance is mainly driven by operational concepts and not the interface design
- nevertheless, a UCCL has the potential to adversely affect mission outcomes if designers do not apply in-depth engineering analysis and careful design practice
- the potential impact on mission performance may force designers to limit the amount of overhead they are willing to incur in implementing a new SoS.

The following are concrete examples of how a UCCL could harm mission performance:

1. In the BMD thread, we found a single interceptor link that is severely bandwidth limited, where any additional overhead may have significant impact on operational performance.
2. In the active protection system thread, we found that some higher-end threats may require a limitation on interface overhead to ensure mission success.

---

[4] *Technical performance* is evaluated using metrics and interface parameters such as delays, data rates, memory use, and data processing time. We note that *interoperability* does not similarly have established metrics or interface parameters that can be used in evaluating these trade-offs, an issue we address in follow-on work to this study.

3. Finally, for the EW thread, we found that even modest inefficiencies may have a significant impact on mission performance.

DoD can mitigate some of the risk associated with implementation of a UCCL by following these general principles:

*   Focus on achieving interoperability for SoS with non-time-critical interfaces or missions with wide performance margins that allow the warfighter to reap the benefits of more or better sensor-shooter pairing. More specifically, efforts should be focused on SoSs (a) that do not have strong dependencies between operational performance and message delays, and (b) whose improved interoperability provides high operational benefits.
*   When evaluating the risk of implementing a UCCL, specifically evaluate and design mitigation strategies for
    -   systems with severely restricted bandwidth links
    -   systems with processors that have very little available processing power
    -   algorithms that require large amounts of data operations per bit of data
    -   systems with submillisecond performance requirements.
*   For interfaces that have tight delay and timing requirements, optimize the interface for compile time composability. In other words, optimize the interface prior to the mission, not dynamically during the mission.
*   Create multiple versions of the standard that are optimized to the performance constraints of the underlying networks and computing nodes. As computing infrastructure becomes more constrained, systems should be able to fall back to less capable but still interoperable versions of the standard for information exchange.

While the modeling we performed in this research was intended to be a high-level investigation of the general trade-offs between interface inefficiencies and operational performance, our methodology could be applied in a more detailed technical analysis of specific systems. Anyone evaluating the applicability of a UCCL to a particular system should

*   translate the mission-critical operational requirements into an interface performance trade-space (e.g., delays versus messaging overhead)
*   quantify the overhead that a specific standard interface implementation would impose

- estimate the impact on operational performance and compare it with operational requirements associated with the mission
- assess whether some of the impact could be mitigated by technical means—for example, by optimizing the particular interface or using an optimized version of the standard. If the issue is related to processing power, then adding more processors could resolve the concern. Where bandwidth is an issue, provide a higher-bandwidth link.
- if the impact cannot be easily or cost-effectively mitigated by technical means, assess whether can it be mitigated through revised tactics, techniques, or procedures. For example, perhaps the warfighter could maintain larger operational distance from the threat, or a change to the force structure might mitigate the threat.
- finally, if the operational performance limitation cannot be mitigated reasonably and cost-effectively, then the interface in question may not be a good candidate for standardization.

We would be remiss if we did not offer one final word of caution regarding DoD's efforts to create a UCCL. While the performance of a standard ensures its technical and operational viability, nontechnical considerations often have a much larger impact on whether it will be broadly accepted in the market (even a constrained market, such as that within DoD). These primarily are related to the economics of standards that can enable ecosystem growth, better retention of human capital, reduced vendor lock, and cheaper training and retraining. To harness these benefits, DoD will need to design and implement an effective and efficient standardization process that addresses and accommodates the interests and motivations of all stakeholders, as well as the legal and regulatory context within which the standardization process will take place.

Acknowledgments

We thank Michael Zatman, Assistant Director, Fully Networked Command, Control and Communications (FNC3), Office of the Under Secretary of Defense for Research and Engineering, for entrusting us with this study and providing invaluable feedback. We thank John Chapin at Roberson and Associates, LLC, for his support of the analysis in his role as action officer, but also for many technical discussions that helped focus and refine our results. Our study benefited from the enthusiastic support and generous participation of numerous individuals in DoD, other federal government organizations and agencies, industry, and academia. Because of our research rules of engagement, we will not identify the individuals we interviewed here. We emphasize that their inputs and perspectives were critical to our study, and we greatly appreciate their time and engagement. Several RAND Corporation colleagues offered their help and support along the way. We appreciate the input from Michael Kennedy, Igor Mikolic-Torreira, Jeff Hagen, and Caolionn O'Connell. We thank Katy Burch-Hudson and Jamila Thompson for helping us organize and host at the RAND facilities a large cross-federally funded research and development center (FFRDC) workshop with multiple government stakeholders without a glitch. We thank our reviewers for their thoughtful feedback: Edward Rutledge from Massachusetts Institute of Technology Lincoln Labs as our external reviewer and Bradley Wilson as our internal RAND reviewer. We thank James Powers, NSRD's research quality assurance manager, for his oversight of the review process, and Saci Detamore as the NSRD Quality Assurance Coordinator for keeping us on track. Finally, we thank the Acquisition and Technology Policy Center leadership team, Joel Predd and Yun Kung, for their enthusiasm and support. While we acknowledge the contributions of all the above, we remain solely responsible for the quality, integrity, and objectivity of our assessment and recommendations.

Abbreviations

AFRL   Air Force Research Laboratory
ANW2   Adaptive Networking Wideband Waveform
APS   active protection system
AWACS   Airborne Early Warning and Control System
BLOS   beyond line of sight
BMD   ballistic missile defense
C2   command and control
C2BMC   Command and Control Battle Management Center
CAC2S   Common Aviation Command and Control System
CAL   critical abstraction layer
CM   cruise missile
COA   course of action
CONCERTO   CONverged Collaborative Elements for RF Task Operations
CPU   central processing unit
CTP   common tactical picture
DARPA   Defense Advanced Research Projects Agency
DDS   Data Distribution Service
DICI   distributed iterate-collapse inversion
DoD   U.S. Department of Defense
DSP   digital signal processing
DT   detection time
ECI   Earth-centered inertial

EO/IR	electro-optical/infrared
ESM	electronic support measures
EW	electronic warfare
FFB	fixed-format binary
FNC3	Fully Networked Command, Control and Communications
FTP	file transfer protocol
GBAD	ground-based air defense
GMD	Ground-Based Midcourse Defense
GPB	general purpose binary
HTTP	hypertext transfer protocol
IBCS	integrated battle command system
ICD	interface control document
IDL	interface definition language
IEC	International Electrotechnical Commission
IFF	identification friend or foe
IFTU	in-flight target update
IHL	internet header length
IoT	internet of things
IP	internet protocol
IPv4	internet protocol version 4
IPv6	internet protocol version 6
ISO	International Organization for Standardization
JREAP	Joint Range Extension Application Protocol
JSON	JavaScript Object Notation
LAAD	low-altitude air defense
LM CAL	Lockheed Martin critical abstraction layer
MADIS	Marine Air Defense Integrated System
MAP	maximum a posteriori
MDD	minimum defeat distance
mips	million instructions per second
MIT	Massachusetts Institute of Technology

MRBM	medium-range ballistic missile
NASA	National Aeronautics and Space Administration
NDRI	National Defense Research Institute
NGMS	Northrop-Grumman Mission Systems
NIEM	National Information Exchange Model
ODN	Open Data Network
OMS	Open Mission Systems
OSI	Open Systems Interconnection
OSPF	open shortest path first
OWL	Web Ontology Language
PEO	Program Executive Officer
P_{kill}	probability of kill
POP	post office protocol
QoS	quality of service
RAM	rockets, artillery, and mortar
RF	radio frequency
RFVM	RF Virtual Machine
RPG	rocket-propelled grenade
RV	reentry vehicle
SBIRS	Space-Based Infrared System
SoS	system of systems
SRT	system reaction time
SSRM	System and Sensor Resource Manager
STITCHES	System-of-Systems Technology Integration Tool Chain for Heterogeneous Electronic Systems
SWaP	size, weight, and power
TA	task activity
TCP	transmission control protocol
THAAD	Terminal High-Altitude Area Defense
TOM	target object map
UAS	unmanned aerial system(s)

UCCL	Universal Command and Control Language
UCI	Universal C2 Interface
UDP	user datagram protocol
UML	Unified Modeling Language
USB	Universal Serial Bus
UTC	Universal Time Coordinated
VMF	Variable Message Format
WiFi	wireless fidelity
XML	Extensible Markup Language
XRCE	eXtremely Resource Constrained Environment

CHAPTER ONE

Introduction

Command and control (C2) of military capabilities have been essential parts of warfighting since the dawn of mankind. However, what began with shouted commands and visual or auditory confirmation that they were being followed has now turned into the challenge of linking, synchronizing, and directing complex heterogeneous systems, under tight time constraints and with the potential of interference by the enemy. The consequences of failure are severe, at both the tactical and operational levels—when systems that do not work well together cannot bring their full capabilities to bear, or even malfunction and cause damage—and at the strategic level—with the loss of battles in wartime and costly acquisition failures in peacetime.

C2 Interface Standardization Challenges

The U.S. Department of Defense (DoD) has therefore undertaken many initiatives over the years to create C2 systems standards for interoperability. Current practices for integration across systems generally rely on the development of Interface Control Documents (ICDs) that describe in detail how the different systems and subsystems connect and interact. While a well-documented and well-understood ICD can ensure that the systems involved exchange the right data in the right format and interpret those data in the right way, ICDs are specific to the systems they are associated with and, therefore, are usually optimized for these systems and their missions. When a new system needs to be integrated, a new ICD is required, and, at some point, the number of ICDs may become unmanageable. Legacy ICDs may also be insufficient for new missions or new combinations of systems. The emerging future vision of joint operations calls for connecting any sensor to any shooter, anytime. If that vision is to be enabled through individual ICDs between all potential sensors and shooters, this would lead to n^2 different interfaces and ICDs, where n is the number of sensors and shooters. Developing, verifying, maintaining, and evolving such a large number of interfaces becomes difficult. Reducing the number of interfaces is known to simplify the design process; reduce costs of design, testing, and maintenance; and increase reliability of the fielded system.[1] Additionally, common interfaces reduce complexity and, therefore, risk, as most problems in systems are commonly found at the interfaces.[2]

Trying to avoid a large number of system-to-system ICDs by instead creating a common interface standard for the system-*of*-systems (SoS) is challenging. If the SoS does not exist yet, assumptions about its emergent characteristics must be made to define appropriate interface elements. If the SoS already exists, additional overhead is created by having to adapt the system components' existing interfaces to the new standard, resulting in reduced performance. In both cases, standardization misses its goal if the standard ends up not being specific or broad enough.

Thus, selecting the right standardization approach and striking the right balance among competing standardization objectives is challenging. The Office of the Under Secretary of Defense for Research and Engineering therefore asked the RAND Corporation's National Defense Research Institute (NDRI) to research key aspects related to the implementation of a Universal Command and Control Language (UCCL) and assess the effects that a UCCL might have on the performance of weapon systems and SoSs. As part of this effort, NDRI researchers reviewed DoD SoS programs to understand their interface requirements, compared those requirements to determine where they fall within the different definitions of common interfaces for DoD and aerospace programs, identified the performance requirements for each of three use cases (mission threads), and assessed how imposing a common UCCL requirement might affect the performance of the systems. We established a conceptual framework for analyzing SoS performance of different sensor-to-shooter connections, combinations, and associated C2 constructs. The intent was not to accurately estimate the performance of the system with and without a universal interface, but rather to explore the range of trade-offs by identifying how performance depends on the characteristics of the standard interface and how it varies with respect to the details of the technical implementation. Consequently, this effort should not be viewed as a study of a specific standard interface but as an early system engineering study of how such an interface could and should be designed. While we recognize that communications and networking effects should be

---

[1]   Ofri Becker, Joseph Ben Asher, and Ilya Ackerman, "A Method for System Interface Reduction Using N2 Charts," *Systems Engineering*, Vol. 3, 2000.

[2]   Olivier L. de Weck, "Fundamentals of Systems Engineering," lecture notes slide deck, Massachusetts Institute of Technology (MIT) Lincoln Labs, Fall 2015.

considered and are included in our models, UCCL itself is intended to be agnostic of the underlying communications systems and networking constructs.

Organization of This Report

The next chapter describes the methodology used by the RAND research team, outlining the approach and documenting caveats and limitations. Chapter Three then presents a discussion of standards for interoperability, followed by a case study on one such standard that illustrates the trade-offs involved and presents important nontechnical considerations. Chapter Four summarizes insights from third-party testing and experimentation based on three different C2 interface standards. Chapters Five through Seven document findings derived from modeling and simulation conducted by the RAND research team for three different DoD mission threads. Chapter Eight provides conclusions and recommendations. A reference section and an appendix containing a short discussion of the origins of the Publish-Subscribe Model and the Data Distribution Service (DDS) are at the end of the report.

CHAPTER TWO

Methodology

The intent of this study is to establish a conceptual framework for analyzing SoS performance of different sensor-to-shooter connections and associated C2 constructs. This framework will connect specific attributes of a UCCL interface, such as message size, with technical performance parameters, such as message delay. We will develop an estimate that relates technical performance parameters to operational performance in the context of a set of mission threads. These estimates of performance will then be used to highlight potential risks that a standard interface might impose on the successful execution of a particular mission.[1]

Approach

The research began with a survey of the existing literature and available experimental or field-testing results to identify performance concerns found in other standardization efforts within DoD and in commercial industry and obtain estimates of the technical performance impacts one might reasonably expect. The intent was to identify specific attributes of interface implementation that affect system performance. Additionally, typical implementation attributes in other standardization efforts will establish a typical range of values for these attributes.

These insights then guided the development of a mathematical model tying technical system performance to specific values for these attributes. We then used the model to estimate technical performance for a range of different values that highlight how different implementations of a standard interface will affect technical system per-

---

[1] It is critical that any analysis like ours be framed within the context of the mission rather than simply in terms of percentage changes of such attributes as mission throughput or latency. Some missions are adversely affected by even a 10-percent change in these factors, while others remain relatively successful even should such parameters double or even triple.

formance. Based on this range of technical performance, we then estimated impacts on the operational performance of the system and the risks to execution of the mission thread. Two different types of impacts were considered: impact on the overall execution of the thread and impact on a critical step in the thread. The critical step analysis was intended to identify any critical steps or exchanges in the thread that might be particularly sensitive to performance overhead and to estimate potential risks in executing this specific step or exchange.

Three mission threads were chosen to highlight a range of different military operations with varying sensitivity to message delays:

- The first was an electronic warfare (EW) mission based on the CONverged Collaborative Elements for RF [Radio Frequency] Task Operations (CONCERTO) system developed for the Defense Advanced Research Projects Agency (DARPA). This is a single system thread with extremely short timelines and very simple C2 exchanges between subsystems.
- The second was an active protection thread modeled on the Marine Air Defense Integrated System (MADIS), in which a few platforms coordinate a defense against incoming threats with relatively short timelines.
- The third was a ballistic missile defense (BMD) thread, with reasonably long mission execution times and a fairly complex C2 construct with multiple sensors, shooters, and C2 elements.

Note that all three of these SoSs have already made a basic trade—the complexity of the C2 is reduced as the needed response times shorten. Yet by studying their implementation in more detail, we hope to learn exactly why and how C2 performance is affected by the performance of the "language" of its interfaces. We found that these three mission threads cover a wide range of scenarios affected differently by underlying technical C2 performance.

For each mission thread, we present the modeling approach, along with the assumptions and limitations specific to the individual thread. The performance is computed over a range of inputs, and the operational performance risks are discussed over this range of potential interface implementation parameters.

Our analysis of each mission thread used the same high-level conceptual C2 construct, shown in Figure 2.1. In this construct, a target operates within an environment, and the mission thread begins with a sensor detecting that target through the environment using an estimation of the physical properties of both the target and its environment. Sensed target and environmental information are sent over a communication link to a C2 authority. The C2 authority uses the information to assess the situation

Figure 2.1
High-Level Mission Threads C2 Model

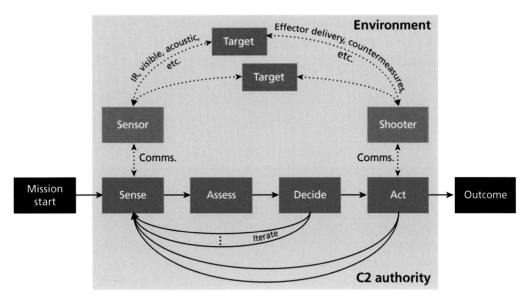

NOTE: IR = infrared; comms. = communications.

and decide on a course of action (COA). If that COA is to engage the target, a message is sent over a communication link to a shooter.[2] The shooter then delivers its effects toward the target. This cycle could iterate several times and involve multiple sensors, shooters, and C2 authorities.

To simplify our analysis, our modeling assumed that there is no human in the loop—i.e., all communication is machine-to-machine, and no human cognition and reaction delays are included.[3] We assumed computing systems at the sensor, C2 authority, and shooter that are capable of appropriately processing the data. We then characterized the effect of delays in data transfer over the communications channel and from processing the data. We included five possible delays from target sensing to the C2

---

[2] The term *shooter* here is used in the broader sense to mean any effect capable of neutralizing an attacker. This can be a kinetic kill vehicle, such as those used to intercept a missile, or an electronic jammer used to bring down electronically controlled vehicles.

[3] In reality, the United States has very few systems that are designed to autonomously make decisions to engage adversary forces. However, were we to include human decision times, they would dominate the overall response times and obscure the mission impacts of alternative interface designs that are the subject of our research. We will return to this limitation of our approach when we discuss the results of our study.

authority and six delays from the C2 authority to the effects reaching the target. The five delays to C2 authority are

1. time to sense the target through the environment
2. time for the computer at the sensor to process the sensed information and create a data packet
3. time to encrypt the packet
4. time to transmit this packet over the communication link to the C2 authority
5. time to decrypt the packet at the C2 authority.

The six delays from C2 authority to reaching the target are

1. time to process the information and create a data packet by the C2 authority computer. We do not include the processing time involved with assessing, deciding, and acting upon the information, which is considered out of scope for this analysis.
2. time to encrypt the packet
3. time to send the packet over the communication link to the shooter
4. time to decrypt the packet at the shooter
5. time for the computer at the shooter to process the information. We do not include any processing time at the shooter to compute a firing solution, which is considered out of scope for this analysis.
6. time for the effects launched by the shooter to reach their target.

Each mission thread model accounted for each of these effects differently. For example, the time it takes a radio frequency signal to travel at the speed of light from the target to the sensor is critical in an EW mission in which small fractions of a millisecond are important. For BMD, where response times are measured in seconds or minutes, these delays can be ignored.

The analysis for each mission thread was also slightly different to account for the different types of operational outcomes each is expected to achieve. A set of assumptions and limitations for each thread is given, detailing how we approached the problem and how one should interpret the results.

We then present the mathematical model used in the analysis of each thread. For the BMD thread, which represents the most complex C2 construct of the three, we developed a model in Python to estimate the end-to-end performance of thread execution. The other two threads are simpler and were modeled as relationships in Excel spreadsheets. The modeling results are then presented, the data interpreted, and we present conclusions.

Finally, we conducted a critical step analysis to identify whether particular steps or information exchanges in the thread are particularly sensitive to performance overhead, and we discuss associated risks.

Assumptions, Constraints, and Limitations

As discussed earlier, the intent of this study was an early system engineering exploration of a generic standard interface. It was not focused on a specific implementation, nor was it intended to estimate the performance of specific implementations or make judgments on specific mission impacts. The analysis drew on unclassified information and available specifications of the underlying systems. Information about detailed subsystem characteristics, such as message sizes or underlying signal processing algorithms, was not available at the unclassified level and was abstracted at a high level in the analysis. As a result, one should not view these results as representing specific systems but rather broad classes of systems like the ones chosen. The results represent an understanding of how performance would vary with respect to varying interface implementations, but not how a specific interface design would affect the specific system in the execution of a particular mission thread.

Additional assumptions and limitations of our modeling efforts were as follows:

- The systems were not assumed to be real-time composable systems. That is, we did not attempt to assess the effects of a C2 system trying to discover different sensors and shooters on the fly. While we are fully cognizant that discovery protocols can have significant impacts on system performance, the performance impact of such a protocol was beyond the scope of this research.
- We made the broad assumption that when adding a sensor-to-shooter pairing to an overall SoS, another sensor-to-shooter pairing is removed. In other words, we did not analyze the impact of allowing all sensors to talk to all shooters, all the time. Any performance impact of adding multiple overlapping message flows was not considered. Additionally, we did not model the process of managing "all sensor, all shooter" data flows or how flows could be added or removed. These effects could substantially affect performance overhead but are beyond the scope of the study.
- We did not include the effect of packet fragmentation. We recognize that all systems have a maximum packet size, and that if a message exceeds that size, it

will be broken into multiple packets. While these multiple packets would add to the overhead associated with sending a message, the effect was not considered because it is a secondary effect and can vary widely, depending on design choices and implementation.

CHAPTER THREE

Standards

Standards to Achieve Interoperability

In this report, we are generally concerned with interoperability standards. For the purposes of this report, we will use the International Organization for Standardization/ International Electrotechnical Commission (ISO/IEC) 25010 definition of *interoperability: "Interoperability is the degree to which two or more systems, products or components can exchange information and use the information that has been exchanged."*[1] As a practical example of what interoperability requires, consider the "any sensor, any shooter" battle management paradigm set forth by DoD's Fully Networked Command, Control, and Communications (FNC3) initiative. To realize that vision requires that sensors and shooters not only have a physical means of communicating the ones and zeros that make up modern information systems—i.e., technical interoperability—but also recognize that those ones and zeros are authorized communications with specific meanings. Ultimately, what is required are standards that allow sensors and shooters to share understanding of the past, present, and future location of targets and authorization to engage selected targets. To achieve this goal, the system must have a shared understanding of the syntax (formats, fields, units of measure) and semantics (definitions of meaning) used in exchanging the information—i.e., syntactical and semantic interoperability.[2]

---

[1] ISO/IEC, "Systems and Software Quality Requirements and Evaluation (SQuaRE)—System and Software Quality Models," ISO/IEC 25010:2011, webpage, 2011.

[2] An example of the difference in syntax and semantics is that syntactically interoperable systems have a shared understanding that position of a target is to be interpreted as meters in three dimensions as opposed to being interpreted as degrees of latitude and longitude and meters of altitude. Systems that are also semantically interoperable know that position of a target is measured relative to a specific reference frame such as Earth Centered Inertial (ECI) as opposed to a frame fixed to the sensor that is providing the target position.

Achieving Technical Interoperability

Over the past 20 years, great strides have been made in achieving technical interoperability of computing systems. A key enabler has been the abstraction of the communication between any two nodes in the system as a *stack* or set of layers that provide the functions required for any two systems to exchange information. The two most commonly referenced abstractions are the Open Systems Interconnection (OSI) and internet models of communication.

The OSI model has seven layers and was designed such that each layer would strictly depend only on the services of the layer immediately below it. Although originally developed with a set of accompanying protocols at each level that would communicate at that layer, the protocols were unwieldy and never caught on. However, the *idea* of the seven OSI layer model persists to this day and provides a common vocabulary for network analysts.[3]

The abstracted internet model is simpler, having only four layers, and is more representative of actual practice. The seven-layer model is contrasted with the four-layer model in Figure 3.1, and commonly used communication standards (i.e., *protocols*) and the media of that communication are roughly mapped against them.[4]

The internet model we show in Figure 3.1, unlike the OSI model, is not formally standardized but instead has become a de facto standard through use. In the same year that the OSI was published (1994), the National Research Council published *Realizing the Information Future: The Internet and Beyond*, proposing a four-layer model that is described thus:

> The Open Data Network proposed in this report involves a four-level layered architecture configured as follows: (1) at the lowest level is an abstract bit-level service, the *bearer service*, which is realized out of the lines, switches, and other elements of networking technology; (2) above this level is the *transport* level, with functionality that transforms the basic bearer service into the proper infrastructure for higher-level applications (as is done in today's Internet by the TCP protocol) and with coding formats to support various kinds of traffic (e.g., voice, video, fax); (3) above the transport level is the *middleware*, with commonly used functions (e.g., file system support, privacy assurance, billing and collection, and network directory services); and (4) at the upper level are the *applications* with which users

---

[3]   ISO/IEC, "Open Systems Interconnection—Basic Reference Model: The Basic Model–Part 1," ISO/IEC 7498-1:1994, webpage, 1994. As of December 2019, it can be purchased online (see references list for URL). However, hundreds and, perhaps, thousands of representations of the model can be found online using the term *OSI 7 layer model* in any search engine.

[4]   Many commonly used protocols have elements of different layers in their makeup.

Figure 3.1
The OSI and Internet Models of Communications

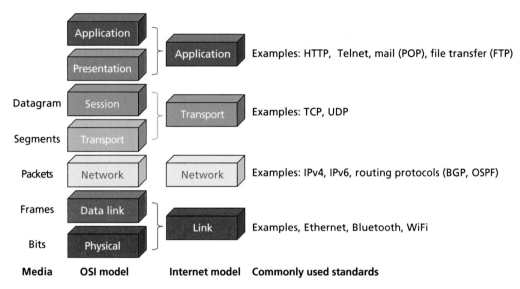

NOTE: HTTP = hypertext transfer protocol; POP = Post Office protocol; FTP = file transfer protocol; TCP = transmission control protocol; UDP = user datagram protocol; IP = internet protocol; IPv4 = internet protocol version 4; IPv6 = internet protocol version 6; BGP = border gateway protocol; OSPF = Open Shortest Path First; WiFi = wireless fidelity.

interact directly. This layered approach with well-defined boundaries permits fair and open competition among providers of all sorts at each of the layers.

In particular, the concept of a distinct bearer service contributes to meeting the key objective of separating the information service provider from the network service provider in order to allow all potential service providers the opportunity to flourish in an ODN [Open Data Network] environment. To provide for this separation, the committee has structured the protocol stack of its architecture such that it narrows down considerably at the interface to the (open) bearer service layer. Above this narrow "waist" the stack broadens out to include the broad range of options for the transport, middleware, and applications layers. Below this narrow waist, the stack again broadens out to include the many possible technologies for implementing network access, local area networks, metropolitan area networks, and wide area networks. Such an arrangement reinforces the principle of separation and is intended not to prevent the same supplier from acting in two roles, but rather to ensure that individual competitors can enter into the marketplace at either level.[5]

Although the names of the layers and the partitions between them morphed in subsequent years, the vision of the internet packet and routing protocols (today's network layer) as being the open standard at the "waist" of a software stack that would permit "fair and open competition" at the layers above and below has been fully realized in the years since. This is an example of a formal, committee-designed standard that was adopted in the market, but the market then adapting it to its various stakeholder needs, thus creating a different common standard with a different structure that achieves the same results.

The ubiquitous use of the internet packet structure and accompanying routing protocols at the network layer of these models is the unifying element that links most communications today. However, this is not true for many weapon systems. Only recently have we been able to package the computing infrastructure needed to support a full IP network stack to meet the constraints of size, weight, and power intrinsic to many weapon systems. Therefore, many military systems currently communicate using highly specialized waveforms; access protocols; and customized, often fixed-format, messages. If two systems use different standards, they can be linked using custom translators. Many of these standards are being updated to be able to carry internet packets, offering the hope of greater technical interoperability in the future, but security and encryption remain difficult issues. Technical interoperability—the proper understanding of bit content and sequencing—achieves only the most basic level of interoperability. Other mechanisms are needed if both sides of the interface are to understand what those data mean.

Achieving Syntactical Interoperability

As we noted earlier, recovering the ones and zeroes that constitute a packet or datagram of information is not enough to ensure weapon systems interoperability—both sender and receiver need a shared understanding of how to convert the transmission into data. The most straightforward way to achieve syntactical interoperability is to assign specific meaning to a fixed group of bits—this is called the *format* of the message. For instance, the binary (i.e., bit-level) format of IPv4 is shown in Figure 3.2. Looking at the format, we know that the four most significant bits of the first word are the version of the packet format, the IP address of the sender is in the fourth word, and the desti-

---

[5]   National Research Council, *Realizing the Information Future: The Internet and Beyond*, Washington, D.C.: The National Academies Press, 1994, p. 5.

Figure 3.2
Bit Structure of an Internet Header

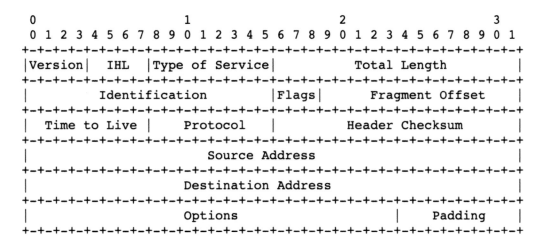

Example Internet Datagram Header

SOURCE: Information Sciences Institute, University of Southern California, *Internet Protocol: DARPA Internet Program Protocol Specification*, Request for Comments 791, Arlington, Va.: Defense Advanced Research Projects Agency, Information Processing Techniques Office, September 1981.

nation address is in the fifth word. But even to understand that format, we first need to know how big a word is (16- and 32-bit words are the usual choices) and the order in which bits and/or bytes are stored or transmitted—some computers expect the first bytes stored and/or received to be the most significant digits, while others expect the first bit or byte received to be the least significant digits (this is called *big endian* and *little endian*).[6] Further complicating things, many software engineers expect counts to begin numbering from zero (for bits, bytes, words, segments, etc.), while non–software engineers usually assume that counts begin at one—meaning that any description of the format that uses bit or word numbering (as opposed to saying "the four most significant bits" or "the fourth word received") is subject to misinterpretation.

A further examination of the packet header shows a common practice in modern computing—*self-definition of interfaces.* The first field tells the receiving nodes the version of the format being used—both IPv4 and IPv6 versions of the protocol are in use today—and by putting this information first, the receiving node can determine whether it has the compatible software stack to interpret the rest of the message cor-

[6] The concept of *endianness* can apply to any sequence of items, but the most common usage of the term is in defining how bytes are transmitted and stored.

rectly. The second field, the *internet header length* (IHL), further defines the format in that it contains a value to tell the receiving nodes how long the header itself is (the standard header is five words, but this four-bit field can hold a maximum value of 15, allowing for a further ten words of optional data). The final field in the first word is the *total length*, defining the total number of words in the packet. Using these two pieces of information, the receiving computer then can determine where the data in the message start (after the last word of the header) and the number of data words contained within the packet (total length minus header length).

So far in this discussion, simply being able to extract the correct bits that define a field and read its binary value has been sufficient to understand the information contained in this header. But other fields are not nearly as straightforward. Consider, for example, the *Protocol* field, which defines what protocol should be used at the next layer up in the software stack. This eight-bit field can hold values from zero to 255, but protocols are rarely referred to by a number, so some other means must be used to convey to the receiving software that a "6" means to handle the message exchange using TCP, while a "17" means to use UDP. This level of syntactical interoperability has typically been achieved through paper specifications that are provided to software developers and then coded into the sending and receiving software items. This paper process is subject to great misinterpretation, even for relatively simple concepts. In one of the better known (and expensive) incidents of misinterpretation of syntax, for a launch vehicle capable of carrying two satellites to orbit, the original designers designated the satellite position with a single bit, meaning the field could be either one or zero. When the design team on one side of the interface interpreted *one* as meaning the upper satellite and the team on the other side interpreted *one* as meaning the lower satellite, the result was a loss of mission.[7]

To mitigate the risk that mistakes in syntax will cause failures in communication between systems, the software industry invented the concept of an *interface definition language* (IDL)—a specification language that provides the syntactical details in

---

[7] This syntactical error led to a wiring error that left the Intelsat VI F-3 satellite stranded in orbit. National Aeronautics and Space Administration (NASA) astronauts flying on the space shuttle later conducted a repair mission. Today, the repair mission is better known than the error that led to it. Wayne Eleazer, "Launch Failures: The 'Oops!' Factor," *Space Review*, January 31, 2011.

Another famous example where lack of syntactical interoperability led to failure in space is that of the Mars Climate Orbiter, which was lost on arrival at Mars. The root cause was found to be that one team had designed using metric units (meters and kilograms) and another had designed using English units (feet and pounds). Douglas Isbell, Mary Hardin, and Joan Underwood, "Mars Climate Orbiter Team Finds Likely Cause of Loss," press release, NASA, 1999.

a language-independent machine-readable format.[8] These IDLs are often associated with a specific set of protocols that form a conceptual *message bus* for the exchange of information across processes, platforms, and systems. Unfortunately, no IDL has become a de facto standard—currently, the Wikipedia page for IDLs lists 22 different languages, with each addressing a different set of features or design challenges (such as performance efficiency). New IDLs are constantly being created as Microsoft, Google, Facebook, and others vie to establish their standard as "the" standard of choice.[9] Even if an SoS wants to use an IDL to define its syntax, it is unlikely that existing systems within the SoS would use the same IDL. Translators are available for some of the most popular languages (for example, Extensible Markup Language [XML] to JavaScript Object Notation [JSON]), but only if the interface design to be translated adheres to the least common set of features between the two languages.[10] While IDLs are invaluable for new software development, incorporating them into an architecture to connect sensors to shooters designed in different eras cannot be achieved in an efficient manner. Features provided by the IDL and its associated protocols in one system will not be supported in other systems, meaning that custom code will need to be created. Often, this is achieved by wrapping the messages associated with the less capable IDL within a message format for the more capable IDL. The net effect can often be nested protocols several layers deep.

Achieving Semantic Interoperability

Semantic interoperability is grounded in a shared understanding of the meaning of data and the relationships between data. For instance, syntactic interoperability ensures that two systems know that the data exchanged represents a price. Semantic interoperability would ensure that both systems understand that the price is in U.S. dollars and does not include tax, and that when associated with an invoice, tax must be included before payment is made.

Ontologies, supported by the Web Ontology Language (OWL) and a smattering of other languages, describe the data of a system and the relationships between them via a formal method. For example, an ontology would show that both price and tax are described by monetary units, that tax in an invoice is calculated based on price

---

[8]   Today, most IDLs can produce interface code for C, C++, Java, Ruby, Python, and many other more-obscure programming languages.

[9]   "Interface Description Language," *Wikipedia* entry, webpage, updated January 4, 2021.

[10]   For instance, JSON supports blank spaces in variable names, but XML does not. Also, XML is a markup language, but JSON is simply a formatter. Therefore, an interface specification that is translatable to both JSON and XML could not use blanks in variable names and cannot require any of the XML markup features.

and location of the sale, and that invoices describe payments due. To fully describe a domain of knowledge typically requires that we define thousands of entities and relationships. Even with the expressiveness of an OWL, semantic interoperability can be elusive.

While development of a foundational ontology on which other ontologies could be based is a subject of ongoing study in software development, domain-specific ontologies have been developed for selected industries. For DoD weapon systems, the most relevant may be the National Information Exchange Model (NIEM) and its predecessor, the Universal Core, which incorporated both "Cursor on Target" and "Situational Awareness" entity-to-entity relationships.[11] The NIEM has been designated as the DoD standard for "vocabulary" since 2013 and is formatted for distribution as an XML schema or using the Unified Modeling Language (UML). The current release is 4.2, which includes a beta tool for building JSON objects from the model.[12]

Interoperability's Impact on Computing Performance

At each layer of the communications model is the concept of a datagram or packet with a header that encapsulates all the bits (headers and data) at the layer above it. This header contains the information needed to interpret, decrypt, route, marshal, or otherwise process the enclosed bits to accomplish the functions allocated to that layer of the architecture. A notional diagram of encapsulating each layer's information on the way down the stack and then stripping it off on the way up the stack to reveal the data transmitted is shown in Figure 3.3. If the datagram or packets can be routed from sender to receiver, and if both sender and receiver have compatible stacks of software, data can be transmitted. Note that the stacks do not need to be identical, only compatible—i.e., they must be able to recognize the header information and process it appropriately to recover the encapsulated information and pass it to the next layer in the stack. At intermediate nodes between senders and receivers, the lower elements of the software stack are used to manage the flow of information using "addresses" and other information contained in the headers.

However, header information is not the only inefficiency introduced by interoperability. At many levels of the stack, information must be exchanged with other nodes to complete the function assigned to that layer of the architecture. For instance, if TCP is used at the transport layer of the architecture, then messages sent via TCP are seg-

---

[11] The situational awareness ontology can be accessed by authorized users online in a document from DoD's Chief Information Officer's Network Management Working Group.

[12] The beta tool for building JSONs is available from the public "Military Operations" domain of NEIM 4.2: National Information Exchange Model, "Movement," database, undated.

Figure 3.3
Encapsulating for Interoperability Increases the Amount of Information That Must Be Passed Between Nodes in a System

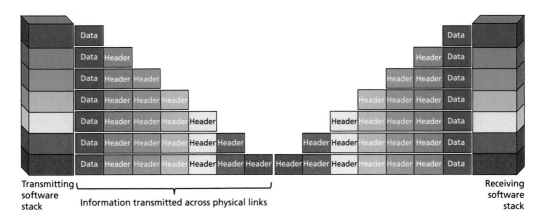

Transmitting software stack

Information transmitted across physical links

Receiving software stack

mented into packets that are numbered before being passed to the IP layer. The receiving TCP processing waits a set amount of time and then requests retransmission of any packets not received within that time. Only when the complete message is reassembled is it passed to the upper layer of the stack. This protocol, while ensuring error-free communication when packets are lost or corrupted in transmission, puts additional messages on the underlying links and can introduce considerable delay in the overall message transmission. At the upper layers of the architecture, where a middleware is used as a virtual message bus, discovery protocols can add a significant number of messages on the underlying physical links—especially when a new node joins the network. The resultant congestion can significantly degrade the near real-time operations needed for targeting in a weapon system. Tactics to control these impacts almost inevitably include a trade-off between latency (delay), throughput (rate of information that can be transmitted), and memory use. A concrete example of those trade-offs is described in the next section.

DDS–eXtremely Resource Constrained Environment—An Example of the Trade-offs Required for Performance

The Data Distribution Service

The eXtremely Resource Constrained Environment (XRCE) extension of the DDS middleware is one of the dominant machine-to-machine communication methods used in real-time systems today. Developed between 2001 and 2004 for defense and aerospace needs, it has become widespread. DDS uses a *publish-subscribe* model to provide efficient and resilient network communications while allowing both the publishers of data and the subscribers to those data to be agnostic to many of the characteristics of the other.[13] In addition to *discoverability*—which allows applications and computing nodes to be added or removed from the network without manual configuration—DDS includes quality of service (QoS) features, which allow network communications to be centrally managed in terms of bandwidth consumed, delivery reliability, and resource limitations. In particular, DDS was designed to provide reliable delivery *at speed and at load*. As a result, the designers of the original DDS made a number of design trades that favored efficiency of transmission and speed of transactions over memory use.

Internet of Things Devices—Beyond DDS's Design Expectations

As networked technologies matured, network-connected devices started to become far more widespread. The *Internet of Things* (IoT) is a term that has come to represent small internet-connected devices that are very lightweight in terms of their processing power and that communicate with larger applications to provide real-time monitoring of items as varied as the status of a pacemaker and the temperature of a home. Because DDS was one of the leading communication standards for machines, developers of IoT devices such as smart thermostats, networked security cameras, and remote health monitors began to use it. However, many of these developers found the performance of DDS incompatible with their needs.

For many applications, the problems arose from the need for the IoT device to run off battery or limited solar power, or the need for the device to fit into an extremely small physical package. The computing hardware is often optimized for extremely low power use and miniaturized for a small footprint. Despite rapid advances in technology that allow us to put greater and greater amounts of memory into smaller and smaller packages, existing DDS implementations consumed too much processing power and memory for many of these devices.[14] While DDS supported discovery to

---

[13] In a publish-subscribe communication paradigm, publishers advertise the data they have available for subscription. This advertisement can include guarantees of the quality of the data provided in terms of staleness, frequency of update, etc. Subscribing services then obtain that data through the middleware. In some publish-subscribe systems the process of matching subscribers to publishers is dynamic and on the fly through a process called discovery. In others, it is more constrained and can even be hard coded. In all cases, publish-subscribe middlewares use IDLs to ensure syntactic interoperability as described in the previous section. A basic primer on publish-subscribe and how it is used in real time systems can be found at PubNub, "What Is Publish-Subscribe (Pub/Sub)?" *Realtime Technology Glossary*, website, undated.

[14] For a recent discussion of the continuing challenge of memory on low-power IoT devices, see Khader Mohammad, Temesghen Tekeste, Baker Mohammad, Hani Saleh, and Mahran Qurran, "Embedded Memory Options

add or remove publishers and subscribers, it did not anticipate large numbers of sensors that would power up from sleep only to report a reading and then disappear from the network again. Because larger sensors deployed alongside very small ones often did support DDS, there was a push to develop a new protocol that would allow all sensors to interact on the DDS network. This protocol became the DDS XRCE specification. DDS-XRCE was able to reach its performance goals only by sacrificing some features. A detailed examination of the trade-offs made may help illuminate factors that must be considered in designing communication systems.

Memory Trade-offs—Losing Memory Flexibility and Gaining Efficiency and Certainty

In many standard DDS implementations, memory is allocated dynamically. Dynamic memory allocation allows the DDS application to use more memory when needed but also to reduce its use when not under heavy load, freeing up space for other applications running on the same computing hardware. However, for an IoT client, dynamic memory becomes a liability because the total memory on the device is extremely small. In fact, dynamic allocation could easily consume all available memory. Therefore, XRCE implementations commonly allocate memory statically, ensuring a specific size of their application memory footprint.[15] While this means the memory cannot be used by other applications even if unused by the DDS-XRCE, the sacrifice ensures that the XRCE cannot consume all available memory. In addition to removing the use of dynamic memory, the developers streamlined the size of the DDS code, allowing the full program for network communications to use less than 75 kb of read-only memory and 2.5 kb of read/write memory.[16] A comparable DDS implementation might require over 100 kb of read/write memory, so the reduction to 2.5 kb is significant.

Bandwidth Trade-offs—Losing Resiliency and Gaining Speed

The DDS-XRCE specification also removes most of the QoS features used in DDS and simplifies the network structure. Instead of full network discovery, microdevices using DDS-XRCE communicate with only a single agent. This agent is typically a computing node permanently connected to the DDS network. The removal of most

---

for Ultra-Low Power IoT Devices," *Microelectronics Journal*, Vol. 93, November 2019.

[15]   Research on secure and reliable memory allocation in IoT devices is a continuing field of study. A recent paper on this topic is Runyu Pan and Gabriel Parmer, "MxU: Towards Predictable, Flexible, and Efficient Memory Access Control for the Secure IoT," *ACM Embedded Computing System*, Vol. 18, No. 5s, Article 103, October 2019.

[16]   eProsima, "Micro XRCE-DDS," webpage, 2019.

QoS functions and routing sections not only affected memory, but also reduced the message size significantly, from 48 to 100 bytes in DDS down to 12 to 16 bytes in XRCE.[17]

However, these changes sacrifice a more resilient mesh network style for a more brittle star network that depends on the single agent computer. The change in network topology means a simpler address field, allowing messages to fit into extremely small memory buffers, but it comes at the expense of network resilience, as described next.[18]

Encryption and Resiliency Trade-offs—Sacrificing Security for Speed

The design decisions above that make the DDS-XCRE bandwidth- and memory-efficient may come at the expense of security or may need to be compromised to achieve security.

The first way the design trades security for efficiency is in the network style. A star network with the agent machine at the center simplifies addressing, routing, and discovery functions in the protocol header, but it also makes the entire network dependent on the single agent computer. If that computer is compromised or destroyed, the entire network ceases to function. A more fully meshed network, with less throughput efficiency, is more resilient to loss of any single node.

We also observe that while it is not a trade-off displayed clearly by XRCE, this switch to a star network and the reduced use of discovery reduce the attack surface for a determined adversary, thus perhaps improving security. Discovery protocols allow computers or sensors to leave the network (for example, to sleep for power consumption, or for maintenance, or because of a power or transmission failure) and rejoin the network later, even if the network configuration (number and addresses of core/agent servers) has changed. In a business context, discovery is enormously useful. However, in a military context, it is often viewed as a security risk and is therefore disabled on operational networks.

Turning our attention to encryption, we note that the small size of DDS-XRCE messages challenges some fundamental aspects of securing the transmission of data. One issue is that encryption algorithms commonly rely on separating transmissions

---

[17]   It should be noted that these optimizations function correctly only if the content of the message is small relative to the header. If systems need to transmit large amounts of per-packet data, the performance of DDS-XRCE is diminished, perhaps becoming unusable.

[18]   Note that because many IoT sensors report only a single item, such as temperature, the standard DDS headers will often be much larger than the information being transmitted. DDS-XRCE is therefore more "efficient" than standard DDS when measured as a ratio of overhead to information transmitted.

into *blocks*.[19] When a transmission cannot be evenly split into packets occupying the full size of the block, the remainder is "padded" with data to confuse attempts to decrypt based on the smaller remainder block. In the case of XRCE, *every* transmission may be quite small, meaning that every transmission may receive significant padding, increasing bandwidth consumption. This extra consumption of bandwidth can be avoided by aggregating data to better fill out the block, but that action might increase latency if all data are not immediately available and creates dependencies that adversely influence later system upgrades.

Another encryption issue that may affect DDS-XRCE applications is rooted in the fact that encrypted traffic can be decrypted more easily when any part of the source text is known. Repeated identical sections of text must be encrypted with a randomized starting value called an *initialization vector*. Large numbers of identical messages mean that the chance of reusing an initialization vector on another message with identical content increases, risking the security of the encryption. Messages that are dominated by header information that is either known or guessable to an attacker, with a very small body of unique information, as is the case with many DDS-XRCE, are therefore fairly easy to decrypt.

However, the ultimate security trade may be decisions to remove encryption entirely when adding small IoT devices to a larger network. While DDS itself often is used with encryption, the processing overhead of encrypting and decrypting all messages would consume significant battery power on a tiny sensor. Given the discussion earlier regarding the encryption vulnerabilities of IoT-generated traffic, the decision to remove it altogether may be reasonable, but the impact on the larger networks of sensors must be considered.

Time Trade-offs

Because lightweight sensors often undergo long sleep cycles, transferring data between the sensors themselves (as DDS allows) was not sustainable: Most of them will be in sleep mode at any given time. This necessitates the use of the agent server as a reliable receiver of messages in a star network, as discussed above. Using an agent computer as a bridge between the normal DDS network and the XRCE network allows the small IoT sensor devices to share their data with the larger network, but at lower transmission speeds than a normal DDS client because the messages need to be passed through the agent and translated into the DDS format from the XRCE format.

---

[19] RSA Laboratories, "FAQ: What Is a Block Cipher?" webpage, 1998.

Summary of Trade-offs

Table 3.1 summarizes the trades that were made by the designers of the DDS-XCRE communications middleware.

DDS-XRCE represents an example of the design trades that must be made to balance memory use, bandwidth use, and speed of communication. It illustrates the need to trade away some attributes to enhance others. It also describes advances being made to push modern communication patterns such as *publish and subscribe* to devices at the very edge of the network. In order to increase bandwidth efficiency and memory footprint, QoS functionality, network flexibility, support for robust security, and encryption were reduced, and dynamic memory management was eliminated. This analysis leads us to observe that a hypothetical UCCL may need to be reconfigurable to perform adequately (or at all) across a wide range of possible environments. If it were not reconfigurable, features important for performance in one task might severely hamper its usefulness in others.

Table 3.1
Protocol Design Trade-offs, According to the Situation in Which Each Returns Benefits

Gain	Sacrifice	Situation Under Which This Trade Provides Benefit
Lower bandwidth use	Reduced routing flexibility, reduced reliability	When header size is large compared with message size
Reduced installation and memory footprint	Loss of dynamic memory allocation; increased effort to implement subsequent upgrades of applications	Useful when total memory and total storage are relatively small compared with application and message size
Increased battery life	Reduced QoS, reduced data protection due to removal of encryption	When encryption is not required, or network load can be preplanned
Reliable messaging when sensors are offline for weeks, months, or years	Less control over message queue size	When traffic volume to edge nodes is small and the agent server is not storage-limited
Allowing sensors to sleep but still send and receive messages when awake	Requires a dedicated agent server to always be present to send and receive—loss of resilience due to no mesh routing capability	When edge nodes in the network will be typically unavailable and do not have spare capacity to handle each other's messages

Nontechnical Considerations for Standardization Efforts

In addition to addressing technology-based challenges, such as balancing flexibility and performance, successful standardization efforts also must overcome nontechnical hurdles. A wide variety of factors may motivate those developing standards. The organizational approach selected to create and implement the standard must address these different motivations to be successful. Both motivations and organizational approaches are discussed in this section.

Why Standards Matter Economically

Foremost among the nontechnical considerations of standardization are the economic effects. The most often stated desired economic effect of standardization is what is called *positive network effects*. A positive effect arises when a social or business network becomes more valuable, and more entities join in, creating a sort of bandwagon effect. In short, each additional standardized platform that is produced provides additional value to each previously converted or produced platform.[20] This may be especially effective in the case of software, as the marginal cost of software distribution is nearly zero, allowing the fixed cost of producing it to be amortized over a larger number of systems or platforms.

Standardization may lower the barriers for entry into a market by thinning the "patent thicket" around some technologies. *Patent thickets* are sets of overlapping intellectual property in which each individual element needs to be negotiated before a product can be produced. Standards-setting organizations, such as the Institute of Electrical and Electronics Engineers, can sometimes help dismantle these thickets by requiring firms to provide fair and equitable access to the patents involved in the standard as part of the standard-creation process.[21]

Transition costs both within firms and within customers (in this case, the government) can decrease as a result of standardization. This decrease is primarily caused by changes in required training and available talent. In an unstandardized environment, labor moving from firm to firm incurs a transitional cost in terms of retraining to use the new firm's set of standards. In a standardized environment, significantly more human capital can be retained when labor moves between firms. Similarly, the end

---

[20] Jeffrey Church and Neil Gandal, "Network Effects, Software Provision, and Standardization," *Journal of Industrial Economics*, Vol. 40, No. 1, March 1992.

[21] Timothy S. Simcoe, Stuart J. H. Graham, and Maryann Feldman, "Competing on Standards? Entrepreneurship, Intellectual Property, and the Platform Paradox," NBER Working Paper 13632, Cambridge, Mass.: National Bureau of Economic Research, November 2007.

consumer can better retain human capital under standardization because changing from one vendor's products to another that follows the same standard is easier, at least in theory.

In the actual production of standardized systems and SoS, another benefit of standardization may be the potential for greater division of labor. Specifically, the increased interoperability that standardization provides may allow firms specializing in particular elements of a system or SoS the opportunity to work in the space where they have a competitive advantage and integrate with each other later in the design process with a lower transaction cost.[22] As a result, vendor lock becomes less of a concern for contracting officers because open standards may allow new vendors to sustain existing systems at lower technical risk.

Why Standardization Efforts Sometimes Fail

As discussed in the previous section, standards matter, and successful standardization efforts have the potential to enable vastly increased capabilities, reduce cost and expenditures, and, thus, ultimately reshape industries.[23] These high stakes can therefore generate opposition from stakeholders that benefit from the status quo, be it via proprietary interfaces or other means of keeping customers captive, or because of increased requirements for user training or other services that these stakeholders may provide, or simply because of increased barriers to entry in their market.[24] Anticompetitive strategies can also include requiring excessive royalties for use of standards-related IP, delaying the reveal of information about IP that is required for the standard, one stakeholder attempting to dominate a standard-setting organization, or outright creating a standard-setting organization beholden to the particular stakeholder.[25]

At the same time, even without active spoilers, standardization efforts can fail because of practical challenges: lack of stakeholder capability, capacity, and/or commitment; failure to involve key stakeholders (including from the user community); per-

---

[22] Carmen Matutes and Pierre Regibeau, "Mix and Match: Product Compatibility Without Network Externalities," *RAND Journal of Economics*, Summer 1988.

[23] Martin C. Libicki, James Schneider, David R. Frelinger, and Anna Slomovic, *Scaffolding the New Web: Standards and Standards Policy for the Digital Economy*, Santa Monica, Calif.: RAND Corporation, MR-1215-OSTP, 2000.

[24] Angelique Augereau, Shane Greenstein, and Marc Rysman, *Coordination Vs. Differentiation in a Standards War: 56k Modems*, NBER Working Paper 10334, Cambridge, Mass.: National Bureau of Economic Research, 2004; Marc Rysman, "Adoption Delay in a Standards War," thesis, Boston University, October 2003.

[25] Jeffrey Mackie-Mason and Janet Netz, "Manipulating Interface Standards as an Anticompetitive Strategy," in S. Greenstein and V. Stango, eds., Standards and Public Policy, Cambridge, UK: Cambridge University Press, July 2007.

sonality conflicts among those involved; legislative or regulatory roadblocks; concerns about IP disclosure and related litigation risk;[26] or standards development timelines that are larger than related technology development cycles and that thus can make a standard obsolete before it is even introduced. The dynamic interaction between hardware and software involved in the systems to be standardized also influences the final equilibrium of the market in which standardization takes place.[27]

In a seminal paper on the topic, Carl Cargill identified the following major categories of standardization effort failures, in order of phase of the standardization process:[28]

1. "The standard fails to get started" (preconceptualization or conceptualization stage)
2. "The standards group fails to achieve consensus and deadlocks" (conceptualization stage)
3. "The standard suffers from feature creep and misses the market opportunity" (discussion or writing stages)
4. "The standard is finished, and the market ignores it" (implementation stage)
5. "The standard is finished, and implementations are incompatible" including because of proprietary extensions (implementation stage)
6. "The standard is accepted and is used to manage the market" e.g., via requiring licensing or royalty fees for use of the standard (implementation stage).

Other researchers have focused on the success factors of standardization efforts and have found that stakeholders' willingness to make technological contributions and their understanding of the relevant market dynamics are correlated with success.[29] Standardization also benefits from network effects: the more users of the standard, the higher the benefits to each user.[30] From the standards user perspective, a recent survey of Air Force Program Executive Officers (PEOs) resulted in a list of "widely accepted common interface standards" that can serve as case studies for successful standards

---

[26] Simcoe, Graham, and Feldman, 2007.

[27] Church and Gandal, 1992.

[28] Carl F. Cargill, "Why Standardization Efforts Fail," *Journal of Electronic Publishing*, Vol. 14, No. 1, Standards, Summer 2011.

[29] Martin B. H. Weiss and Marvin Sirbu, "Technological Choice in Voluntary Standards Committees: An Empirical Analysis," *Economics of Innovation and New Technology*, Vol. 1–2, 1990.

[30] Michael Katz and Carl Shapiro, "Systems Competition and Network Effects," *Journal of Economic Perspectives*, Vol. 8, No. 2, Spring 1994.

development.[31] The following guidelines, which are based on these insights, can help avoid standardization failure.

1.   Design and implement an effective and efficient standardization process, which involves:
     – including the appropriate stakeholders, including future users of the standard
     – enabling a process timeline that is faster than that of underlying technologies and markets
     – transparent deliberations and decisionmaking
     – clear rules for resolving disagreements and overcoming deadlock
     – considering an implementation strategy from the beginning
     – minimizing costs to the participants of the standardization process and to users of the standard.

2.   Understand stakeholders and their motivations:
     – Anticipate stakeholder self-interest.
     – Mitigate against undue influence by a single stakeholder or group of aligned stakeholders.
     – Ensure that the standard provides clear benefits, not just for the market as a whole, but also for all stakeholders.

3.   Understand the context in which the standardization process takes place:
     – legal and regulatory constraints and opportunities
     – intellectual property considerations and strategies
     – related products, services, markets, and associated standards
     – network effects.

Conclusions

Implementation of standards to achieve interoperability will require the design team to trade off various aspects of computing system performance. These trade-offs involve such performance parameters as delays, data rates, memory use, and data processing. Adding interoperability concepts to machine-to-machine interfaces inevitably adds headers to each data exchange and may require that additional data exchanges take

---

[31]   Stephen J. Falcone, "Modular Open Systems Architecture/Approaches," briefing slides, U.S. Air Force Life Cycle Maintenance Center Battle Management Directorate, February 6, 2018.

place. To extend the standard to small devices at the tactical edge may require sacrificing highly desirable attributes of a communications system, such as composability, resilience, or even security.

While the economics of standards may enable network effects, better retention of human capital, reduced vendor lock, and cheaper training, nontechnical considerations often result in standards that fail to gain widespread acceptance in the market. Best practices to avoid such a fate include designing and implementing an effective and efficient standardization process; understanding stakeholders and their motivations; and understanding the market, legal, and regulatory context within which the standardization process will take place.

With this background established, we next turn our attention to understanding how existing test data may give us insight into performance impacts of different interface standard implementations.

CHAPTER FOUR

Insights from Testing and Experimentation

In many areas, standards are specific to a narrow range of applications, which allows them to be customized for increased efficiency and reduced complexity. For example, peripherals are connected to a computer using a different standard (e.g., Universal Serial Bus [USB]) than the one for connecting computers to each other (e.g., Ethernet), and short-range wireless communications use a different standard (e.g., Bluetooth) than do medium-range ones (e.g., WiFi).[1]

Developing a single standard to cover the connectivity needs of a wide range of systems and SoSs, as is the objective of the UCCL effort, is expected to make establishing C2 links between those nodes more efficient, flexible, and robust. However, there are potential disadvantages, such as the additional overhead, transmission delays, or complexity that can be introduced by a one-size-fits-all standard.

Testing and experimentation can provide insights into how to strike the right balance when developing and implementing a universal standard for C2 links. This chapter therefore reviews results from experimental performance testing and comparison of the SoS Technology Integration Tool Chain for Heterogeneous Electronic Systems (STITCHES), Open Mission Systems/Universal C2 Interface (OMS[2]/UCI),[3] DDS,[4] and Variable Message Format (VMF)[5] standards conducted by other parties. The intent is to identify real-life quantitative examples of performance impacts from differing implementation of standards and how widely they vary. We also identified real-life examples of message sizes resulting from differences in standards. This infor-

---

[1] Even this statement alone understates the specificity of standards—Bluetooth, for example, is not a single standard but a suite of standards (Bluetooth SIG, "Specifications," webpage, 2021).

[2] U.S. Air Force, "Open Mission Systems (OMS)," briefing, September 27, 2017b.

[3] U.S. Air Force, "Universal C2 Interface, Part of the Open Mission Systems," briefing, 2017a.

[4] Object Management Group, "Data Distribution Service Specification Version 1.4," spec sheet, March 2015.

[5] SyntheSys Defence, "Variable Message Format (VMF)," fact sheet, undated.

mation informs the modeling in subsequent chapters and provides some comparison performance data to ensure modeling outputs are comparable to real-life results.

STITCHES Performance

STITCHES is an approach for connecting heterogeneous systems by establishing message interface standards between each system and an adjacent system, with messages between systems that are not directly connected being successively translated by each of the systems in the transmission chain. A DoD contractor recently undertook a series of tests in which STITCHES was integrated in different ways with the Lockheed Martin Critical Abstraction Layer (LM CAL) for OMS to connect subsystems, and the differences in performance were measured. The different systems were implemented on an Intel i7 processor with 4 Cores/8 Threads @ 2.2 GHz (to 3.20 GHz Turbo Frequency) and 16 GB RAM. Each subsystem was simulated on a different processor core, and messages were exchanged between cores. Messages could be sent in the native format (unpacked) or in an optimized smaller size format (packed).

Based on the interviews with engineers supporting the programs and a review of results,[6] typical processing delays (excluding communications and networking delays) for these different standard interface implementations are on the order of milliseconds on a circa 2018 computer processor. The details of the implementation of the standard seem to affect the delay by an order of magnitude. The impact on the processor overhead was less clear because of the way it was measured, but there seemed to be significant differences in the speed at which certain standard encoding and decoding could be processed. This finding indicates that any investigation of interface performance should account for both delays and the processing overhead.

OMS Performance

OMS is one of the U.S. Air Force's *Open Architecture Management Standards* that "enable current, legacy, and new programs to realize the benefits of open architecture."[7] OMS facilitates the integration of new subsystems in U.S. Air Force aircraft and other major platforms and enables communication among them. The Air Force Research

---

[6]   The results we reviewed are not available to the general public.

[7]   U.S. Air Force, 2017b.

Laboratory (AFRL) conducted a performance characterization of OMS that looked at the impact of different implementations of the OMS Critical Abstraction Layer (CAL).

The messages were first serialized, i.e., translated in a more efficient *on-the-wire* format for transferring over a communication link, and then deserialized, i.e., translated back into readable format. Performance metrics tracked included average processing times across all OMS messages used in the testing, as well as the average message size after processing by the CAL. It was observed that overall delays can vary by an order of magnitude, depending on the implementation of the standard. Therefore, any investigation of interface performance must account not only for the changes in resulting sizes of the messages, but also for the higher complexity associated with processing the different messaging standards.

Data Distribution Service Performance

DDS specifies a "Data-Centric Publish-Subscribe model for distributed application communication and integration" that facilitates efficient dissemination of information among heterogeneous systems.[8] A study undertaken at Vanderbilt University[9] investigated the performance of three different architectural implementations of the DDS standard:

1. *decentralized architecture*, in which user processes in the connected systems communicate directly with each other
2. *federated architecture*, in which system-to-system communication is handled through a separate process in each connected system that interfaces with the system's user process
3. *centralized architecture*, in which a dedicated system, running a dedicated process, manages the message traffic among participating systems (but participating systems still transfer data directly to each other, once the message manager has initialized communications).

Testing took place on dual-processor blade computers connected through a Gigabit network. Performance metrics again included latency and throughput, as well as

---

[8]   Object Management Group, 2015.

[9]   Ming Xiong, Jeff Parsons, James Edmondson, Hieu Nguyen, and Douglas C. Schmidt, *Evaluating the Performance of Publish/Subscribe Platforms for Information Management in Distributed Real-Time and Embedded Systems*, Nashville, Tenn.: Vanderbilt University, 2011.

the standard deviation of the latency (*jitter*). Table 4.1 and Table 4.2 show the results.[10] It is evident that the DDS-1 architecture provided better performance than did DDS-2 and DDS-3, as measured both by latency and jitter and by throughput. Furthermore, the multicast communications model offered much better bandwidth,[11] making the DDS-1/multicast combination the best performer.

This experiment also looked at scalability, i.e., how the performance metrics are affected by the number of systems communicating. Figure 4.1 shows how both the number of subscribers for a message and the size of the message affect bandwidth for the DDS-1 implementation and the multicast communications model:[12]

- Increasing the message size reduces the impact of overhead and thus increases the effective bandwidth.
- Increasing the number of subscribers significantly reduces relative performance for unicast, but not for multicast.

Table 4.1
Results of DDS Architecture Investigations as Reported by Xiong et al., 2011: Latency and Jitter, by Implementation

UNCLASSIFIED

Implementation[a]	Architecture	Roundtrip Latency (ms) ↓	Jitter (μs) ↓
DDS-1	Decentralized	0.083	4
DDS-2	Federated	0.223	10
DDS-3	Centralized	0.433	23

NOTES: Results shown are approximate, for a message length of 64 samples, simple data type, 1-to-1 messaging on the same blade. ↓ = lower is better; ms = milliseconds; μs = microseconds.
[a] As per Xiong et al., 2011 (Figures 7 and 8).

---

[10] All values shown in these tables were extracted from the figures in Xiong et al., 2011, using WebPlotDigitizer (Ankit Rohatgi, WebPlotDigitizer, Version 4.4, web-based tool, November 28, 2020).

[11] The authors of the DDS study define bandwidth as the maximum kilobits of data that can be sent in one second using a given communication model and network architecture. Therefore, a higher bandwidth score indicates a more efficient use of the available bandwidth on a link; i.e., fewer overhead bits are needed per data exchange.

[12] DDS-1 multicast was chosen because it is the highest-performing combination, according to the results shown in Table 4.2 and Figure 4.2.

Table 4.2
Results of DDS Architecture Investigations as Reported by Xiong et al., 2011: Bandwidth, by Implementation and Communications Model

Implementation[a]	Architecture	Bandwidth (kb/second) ↑		
		Unicast[b]	Multicast[c]	Broadcast[d]
DDS-1	Decentralized	1,320	10,700	
DDS-2	Federated		1,860	1,120
DDS-3	Centralized	930		

NOTES: Results shown are approximate, for a message length of 256 bytes and 12 subscribers. Blank cells indicate that the communications model is not supported by this implementation. ↑ = higher is better.

[a] As per Xiong et al., 2011 (Figures 9, 10, and 12).

[b] Sender addresses message to one recipient.

[c] Sender addresses message to multiple recipients.

[d] Sender transmits message to all connected recipients.

Figure 4.1
DDS-1 Scalability

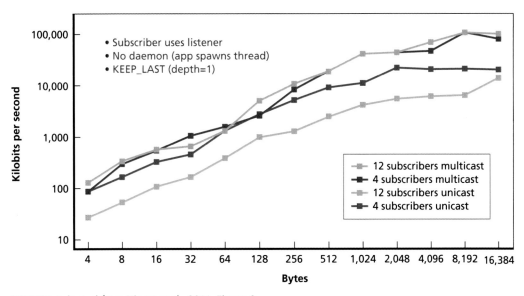

SOURCE: Adapted from Xiong et al., 2011, Figure 9.

We see in these results that even if the same interface standard is used, the way we architect the connections as well as the number of systems participating in the communication can have an order of magnitude impact on performance.

Analyzing Universal Command and Control Interface and Variable Message Format

The UCI is a message interface standard that is mandatory for use with new U.S. Air Force acquisition programs.[13] It also has been incorporated into multiple existing platforms. VMF is a message interface standard used in U.S. Army systems.[14] Massachusetts Institute of Technology (MIT) Lincoln Laboratory analyzed and compared the performance of these two standards and assessed the performance impact of translating from one to the other during real-time message exchanges.[15] Two application cases were modeled: a system to system exchange of data where U.S. Air Force sensors are used to generate target data for a U.S. Army artillery shooter and an internal exchange of data between sensors and shooters integrated into a ground vehicle's active protection system. In the system to system exchange, the Air Force systems use the UCI for their internal communication and the Army system uses VMF, requiring a translation step to be inserted between the two systems. The UCI implementation included a human-readable XML format, compressed XML format, and a general-purpose binary format based on a Google Protocol Buffers implementation. The Army VMF implementation also included the standard Army fixed-format binary (FFB). For the VMF case, the resulting message sizes were 2,577 bytes for XML, 785 bytes for compressed XML, 115 bytes for general-purpose binary, and 47 bytes for FFB. In this study, there was no effort made to optimize the schema (for example, in the compressed XML case a simple and nonoptimal gzip operation was used for the compression), and the now-standard and more optimal VMF XML schema was not used. Performance metrics used were latency, throughput, bandwidth requirements, central processing unit (CPU) load, and memory consumption. This work produced several insights:

---

[13] "Use of Open Mission Systems/Universal Command and Control Interface," Department of the Air Force memorandum for Air Force PEOs, October 9, 2018.

[14] SyntheSys Defence, undated.

[15] Edward Rutledge et al., *UCCL Performance Study Report*, Cambridge, Mass.: Massachusetts Institute of Technology Lincoln Labs, forthcoming.

- Translating one message from the source standard may result in multiple messages in the destination standard, and vice versa.
- Different encoding options for messages result in different performance characteristics (code complexity, human-readability, processing effort, message size).
- On fast transport networks (e.g., Gigabit Ethernet), message processing and transmit times are substantially below the upper limit imposed by even the challenging APS application case (challenging because it calls for a response time on the order of magnitude of milliseconds), independent of message encoding and resulting message size.
- However, on bandwidth-limited transportation networks (e.g., Link-16), the message size, which is driven by the encoding option, can lead to transmission times on the order of magnitude of several seconds per set of messages, which is beyond what is acceptable for the APS case. Thus, encoding for time-critical applications must be optimized for message compactness rather than readability.

Potential Inefficiencies of a UCCL

From these past results, we observe that different implementations of different standards can have significant impacts on performance. The most pronounced effect is on system delays. Factors that influence delays are message size, data processing complexity, transfer time over a communication link, and how the messages flow from system to system. Our ability to discriminate between performance of different interface implementations was dependent on our ability to capture the effect of these parameters adequately.

Message size, as discussed in Chapter Three, can grow quickly if multiple layers of communication are used to complete a message exchange. An interface designed to operate exclusively between two specific systems can be optimized to exchange the minimum amount of information needed to execute the mission. If these exchanges are tailored to specific systems and for a specific set of actions or missions, a clever designer can tailor a very small message set and associated data fields without the need to exchange any information about the systems themselves, the mission, or underlying information that does not change over time or across missions. However, if the systems are designed to allow multiple sensors to connect to multiple shooters in ways that vary over time, in support of a wider variety of missions, then additional information needs to be exchanged between the systems. For example, when a new sensor connects to a new shooter, they need a handshake to identify themselves, their respective capabilities, and the mission they are executing. This requires additional data headers, data fields, and a negotiation protocol to agree upon a basic set of rules about their mission and the data to be exchanged during mission execution. These headers play a role like the IP headers described in Chapter Three, enabling interoperability while abstracting the underlying systems complexities.

Another inefficiency introduced by a UCCL may be the need for larger data fields. All systems are designed to use data at a particular precision that allows their algorithms to make accurate calculations and decisions within the context of the mission. If only two systems are connected, then message data fields can be sized to the needs of each system and algorithm. When multiple sensors need to connect to multiple shooters, data fields are often sized to the system that requires the highest-precision inputs. So, while two systems could execute the mission by passing a data element as an integer variable, the overall SoS might require a single or double precision variable to support a third system that needs higher accuracy.[16]

These inefficiencies will create extra delays because of additional processing time, additional encryption and decryption time, and additional transmission time over the communication links. The processing time will depend on the speed of the underlying processor and the complexity of the operations to be performed on the additional header and data fields. The encryption and decryption time will depend on the speed of the underlying cryptographic device and additional amount of header and data that need to be encrypted/decrypted. The transmission time will depend on the data rate of the communication link and the additional amount of header and data that need to be exchanged.

Summary of Testing and Experiment Data Review

From the cases discussed earlier, it is evident that performance—as measured by throughput or bandwidth and latency—is greatly affected by implementation details.

The STITCHES-based approach, especially when combined with sender-packed messages, shows substantial differences in performance, depending on the configuration. The details of the implementation can affect the delay by an order of magnitude.

---

[16] Obviously, there is a trade to be made here. Many languages support *self-describing* interfaces—an approach that uses extra headers and protocols to describe the data formats used by the system. This approach allows data fields to be sized to only what is needed by the sending or receiving system but comes at the expense of the extra messages, headers, or data fields needed to describe the interface.

The impact of the additional CPU processing can be two orders of magnitude. Therefore, both delay and processing overhead are important.

In the case of the OMS experimentation, the reference CAL is significantly slower and less efficient regarding the message size it produces. Again, this confirms that implementation details matter greatly for the performance of C2 standards and can vary by an order of magnitude. Additionally, the higher complexity associated with processing some messaging standards can have a significant effect.

The DDS experiment illustrates that implementation details—such as architecture, communications model, and number of participants—can have significant impacts on system performance, even when the standard used is the same.

Finally, an analysis of the UCI and VMF standards for two different use cases again emphasizes that performance is greatly affected by such implementation details as message encoding and compression and depends on the underlying transmission protocol used and the speed of the communication link. We also saw an example of how different implementations of a standard can result in different packet sizes in real-life implementations. These include FFB, general-purpose binary, compressed XML, and regular XML. We will use these data sizes as proxies in the following chapters for different interface implementations that will include modeling of specific mission threads and system implementations.

In Chapters Five through Seven, we analyze three different mission threads:

- EW inter-pulse jamming
- vehicle APS
- BMD.

We chose these threads to represent a wide variety of execution times, from sub-milliseconds for an EW mission to milliseconds for an APS mission to minutes for a BMD mission. For each of these threads, we modeled the effects of delays on the operational mission effectiveness and identified the risks associated with different levels of implementation overhead.

CHAPTER FIVE

Electronic Warfare Mission Thread

CONCERTO is a program designed to reduce size, weight, and power (SWaP) concerns in military aircraft. It does so by enabling sharing of RF hardware for onboard systems such as EW, communications, and radar. Developed for DARPA, CONCERTO relies on modern, highly capable RF equipment paired with novel methods of implementation. This makes CONCERTO highly mission-adaptable: Antennae could be added or removed without changing the system interface, and aircraft with smaller payloads can become more capable of performing multimodal missions. By using flexible common hardware, SWaP can be greatly reduced, so long as the hardware can switch between functions in a sufficiently timely manner.[1] In our analysis, we consider a genericized version of the CONCERTO system concept. The system we consider must have three important parts:

1. An antenna, which receives RF and provides digitized RF to the Common Hardware. This antenna is shared by all the possible RF functions.
2. Common processing hardware, which configures the antenna and processes data. This processing hardware and the antenna being common between RF functions is what allows the reduction in SWaP.
3. An adjudicator, which manages common processing hardware and antenna use. This component must exist to adjudicate use of the other two pieces of the system.

In this chapter, we explore how a UCCL implementation might interact with a system that implements a CONCERTO-like system concept, hereafter referred to as the System. The mission thread we explore is an EW inter-pulse jamming mission, which requires extremely short submillisecond reaction times, for which interface over-

---

[1] Kevin Rudd, "CONverged Collaborative Elements for RF Task Operations (CONCERTO)," webpage, Defense Advanced Research Projects Agency, undated.

head delays will have a more pronounced effect on mission performance than will be observed in the mission threads discussed in Chapters Six and Seven.

Modeling the Mission Thread and Its Interfaces

We must consider the level within the System's process at which to insert UCCL. Within a subsystem, whether an antenna, subsystem control, or resource management, UCCL offers little benefit for potentially significant performance impacts. Data within these subsystems are rapidly exchanged, meaning that any excess header information leads to compounding delays. Furthermore, these subsystems are mostly self-contained, with only very specific inputs and outputs. As a result, it makes more sense to imagine UCCL as the method through which these subsystems communicate, rather than a language internal to any or all of them. In this way, UCCL assists in CONCERTO's goal of adaptability and openness.

We examine a potential use case in which an antenna aboard some platform, hereafter referred to as the CONCERTO platform, is being shared between Electronic Support Measures (ESM) and EW jamming activities. In this scenario, outlined in Figure 5.1, an airborne platform in ESM mode detects a communications signal that it intends to jam.

After detection, the common processing hardware and adjudicator work to prepare and manage the switchover of the antenna's use from ESM to EW mode. We examined an approximation of this process and explored how different potential UCCL headers affect the System's ability to jam the pulse in terms of the distance over which this can be done. This distance, d_{jam}, is defined by the time available for the signal to propagate, which, itself, depends on the pulse of the signal to be jammed and the time to move and process information within the System.

$$d_{jam} = \frac{c}{2}\left((1-f) \times t_{pulse} - t_{internal} - t_{xfer}\right). \qquad 5.1$$

In Equation 5.1, we let c be the speed at which the signals propagate (assumed to be the speed of light); f is the portion of the signal that must be jammed to be considered a success; and t_{pulse}, $t_{internal}$, and t_{xfer} are the pulse length, subsystem internal processing activity time, and data transfer and header processing time, respectively. We can see that the time remaining after accounting for the amount of the pulse that must be jammed and information transfer and processing time gives us the time for the signal

Figure 5.1
Electronic Warfare Mission Thread High-Level Operational Concept

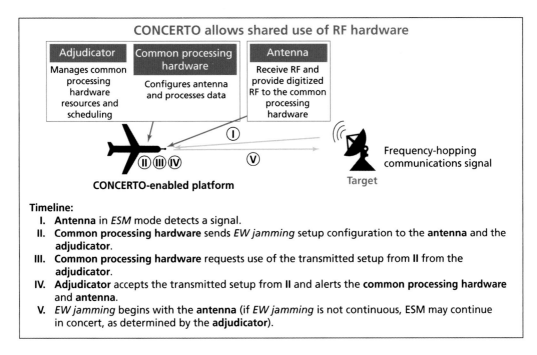

Timeline:
   I. **Antenna** in *ESM* mode detects a signal.
  II. **Common processing hardware** sends *EW jamming* setup configuration to the **antenna** and the **adjudicator**.
 III. **Common processing hardware** requests use of the transmitted setup from **II** from the **adjudicator**.
  IV. **Adjudicator** accepts the transmitted setup from **II** and alerts the **common processing hardware** and **antenna**.
   V. *EW jamming* begins with the **antenna** (if *EW jamming* is not continuous, ESM may continue in concert, as determined by the **adjudicator**).

to propagate to the receiver and then from the jammer to the target. And knowing the speed of the signal, we can determine d_{jam}. We derive this equation later in this chapter (see derivation for Equation 5.4, later). UCCL will come into play by potentially altering time t_{xfer}.

The System Interfaces

As mentioned previously, the System revolves around the interaction of three key pieces: the antenna, the common processing hardware, and the adjudicator. The process by which these parts interact in our example scenario is detailed here.

The process begins with the common processing hardware recognizing the ESM reading as a signal to jam and sending instructions to switch the antenna from ESM to EW mode. These instructions are sent to the adjudicator (to ask permission to do the switch) and the antenna (to do the switch). The adjudicator determines appropriate use of the antenna and enables the common processing hardware's request. The antenna is configured into EW mode while the common hardware provides the EW waveform and jamming begins.

As mentioned earlier, we explored the impact of UCCL as a requirement for interaction between the antenna, common processing hardware, and adjudicator but did not consider the use of UCCL within these subsystems and view each as being effectively a black box provided by a manufacturer.

Unfortunately, because of this necessary black box approach, we do not know the internal processing time to accomplish tasks within the antenna, common processing hardware, or adjudicator subsystems and were forced to estimate them. Consequently, it was difficult to fully simulate the process described above. Instead, we looked for the most likely *long pole*, the interactions that most likely dictate how quickly the process happens. Our candidates for these are the following:

- the common processing hardware sending the setup data to the antenna, receiving the status of that setup back from the antenna, and sending the waveform data to the antenna
- the common processing hardware sending the setup data to the adjudicator, sending an access request to use the antenna for EW purposes to the adjudicator, and the adjudicator enabling the configuration for the antenna.

If we assume that the processing time in the subsystems is similar between the two candidates, then the second candidate is likely to take longer, because the I/Q data are smaller than the control message, and both transport over hardwire connections with similar throughput. This is the path that we modeled to compare the effect of possible UCCL implementation.

The Model of the UCCL Interfaces

Our notional model includes the adjudicator, common processing hardware, and antenna each as nodes. UCCL would be implemented for each in such a way that their inputs and outputs are compatible with a UCCL. This means that all steps in Figure 5.1, with the exception of the antenna's RF transmission, are candidates for using UCCL. As mentioned previously, we examined the likely longest path and explored the impact of UCCL across these three steps:

1. The antenna setup message is sent from the common processing hardware to the adjudicator.
2. The access request is sent from the common processing hardware to the adjudicator.
3. The access configuration enabling message is sent from the adjudicator to the antenna.

Because these steps happen sequentially, we could calculate their time to occur independently and sum them to get the total time. The internal processing time for these steps—that is, the time to handle internal data processes not having to do with interpreting received data—is dependent on the specific equipment and design of the hardware and software and was unknown to us. As this was outside our concern, we abstracted these times to a fixed value, $t_{internal}$, that represents the summation of the time for all internal processing.

We calculated the time to complete a data transfer and interpretation step, n, as

$$t_{xfer_n} = \frac{packet\ size}{transfer\ speed} + packet\ size\ \frac{operations\ needed\ per\ bit}{processor\ speed}. \qquad 5.2$$

By summing these individual times for the traditional and possible UCCL packet sizes, we could determine the time difference between these options:

$$t_{xfer} = \sum_n t_{xfer_n}. \qquad 5.3$$

If we wanted to jam some percentage of the communications signal, f, as mentioned previously, then we could conclude that the time left after accounting for this percentage, the internal processing time, and t_{xfer} is the time left for the signal to propagate (first from the source to the System platform, then from the System platform to its target). And the time to propagate is the distance traveled divided by the speed of light, at which the RF signal propagates (the maximum jammable distance). This gives us Equation 5.4, which we can rearrange to derive Equation 5.1.

$$2 \times \frac{d_{jam}}{c} = (1 - f) \times t_{pulse} - t_{internal} - t_{xfer}. \qquad 5.4$$

Modeling Assumptions

The notional signal to be jammed is a frequency-hopping communications signal with a pulse length of 1 ms. It is commonly assumed that successful jamming requires the ability to jam at least one-third of the pulse.[2]

We assumed that the System platform is equidistant from the source of the communications signal and its intended point of jamming. Because we could not know the exact geometry of this setup, we chose this simple representation that allows easy

---

[2] Richard Poisel, *Modern Communications Jamming: Principles and Techniques*, Washington, D.C.: Artech House, 2011.

comparison between our cases. To keep this simple geometry, we also did not account for issues related to factors such as curvature of the Earth and, instead, envisioned an infinitely flat expanse over which we compared our cases.

Because we could not be sure of the specific capabilities, we further assumed that the time to process information and move from one communications step to the next for the antenna, common processing hardware, and adjudicator is similar. We estimate the sum total of this time for the process we investigate as 0.1 ms.

Table 5.1 lists the modeling parameters and assumptions.

Modeling Results and Critical Step Analysis

The analysis encompasses the entire intra-pulse jamming sequence, because it is composed of a small number of rapid steps, and all need to be accomplished in a specific order for successful jamming. In the context of this mission thread, each step is critical. We explored four different UCCL candidates: an FFB 47-byte UCCL, a general-purpose binary 115-byte UCCL, a compressed XML 785-byte UCCL, and an uncompressed XML 2,577-byte UCCL. These should not be interpreted as estimates

Table 5.1
Modeling Parameters

Parameter	Units	Value
Internal "black box'" processing time for all systems combined	ms	0.1
Message size–baseline	bytes	4 or 16
Message size–UCCL	bytes	47, 115, 785, or 2,577
Number of operations required per bit	—	200 to 10,000
Processor speed	mips	10,000 to 300,000
Pulse length	ms	1
Ratio of distance to source versus distance to target	—	1
Signal propagation speed	m/s	3×10^8
Successful jam ratio	—	1/3
Wired bandwidth	GbE	1 or 40

NOTE: GbE = Gigabit Ethernet; MIPS = million instructions per second; ms = millisecond(s); m/s = meters per second.

of overhead for each implementation approach within the System, but rather as a range of possible input values. Implementations of each approach will differ and depend on several design decisions that are beyond the scope of this study. Hence, when we reference performance of an XML implementation as compared with an FFB, we are referring to these representative values of message sizes and not actual XML or FFB implementations of a standard. We then compare these with an assumed base case of header sizes of around 5–25 bytes. We then varied the processor speed from 10,000 to 300,000 mips and the number of operations needed per bit from 200 to 10,000. Note that while we do not have estimates of the operations per bit required for each format, in general it would be increasing going from FFB, general purpose, to XML, to compressed XML. Figure 5.2 showcases these results.

A typical IP header processing complexity factor is around 100 operations per bit, while typical signal processing complexity factors are 10,000 operations per bit or more.[3] We explored the range of processing complexity factor up to 10,000, assuming that the processing algorithms must have been designed with maximum computational efficiency in mind. These results should be used to understand the trade space for these types of systems, and not as performance numbers for a specific system. In our notional example and under the assumptions stated above, the System platform can jam a communications signal at a distance of at least 50–60 km for almost all ranges of values for processor speed and required number of operations per bit. Indeed, in many of the cases we explored, the jammable distance was greater than 80 km. An FFB 47-byte value may reduce this distance by anywhere from 10 to 30 km in most cases, and more in select cases where a large number of operations per bit is required or processor speed (or available additional processing power) is very low. In the most extreme of these cases, the system fails to complete its objective. The general-purpose binary value of 115 bytes continues this trend, with the jamming being impossible because of delays in transferring and interpreting data for a number of cases in which the number of operations per bit is high and processing speed is low. Finally, the compressed and uncompressed XML values make this mission impossible except in cases in which we need very few operations per bit and have a very fast processor—and in most of those cases, our performance is still severely degraded compared to the base case.

The source of this degradation must come from one of two sources: the time required to transport the data or the time required to process and interpret them once received. Because the communication links in this example are rated at more than 1 GbE (due to our assumption that they are hardwired links), we can conclude that they

---

[3]   R. S. Tucker and K. Hinton, "Energy Consumption and Energy Density in Optical and Electronic Signal Processing," *IEEE Photonics*, Vol. 3, 2011.

Figure 5.2
UCCL Integration Results: Maximum Jammable Distance

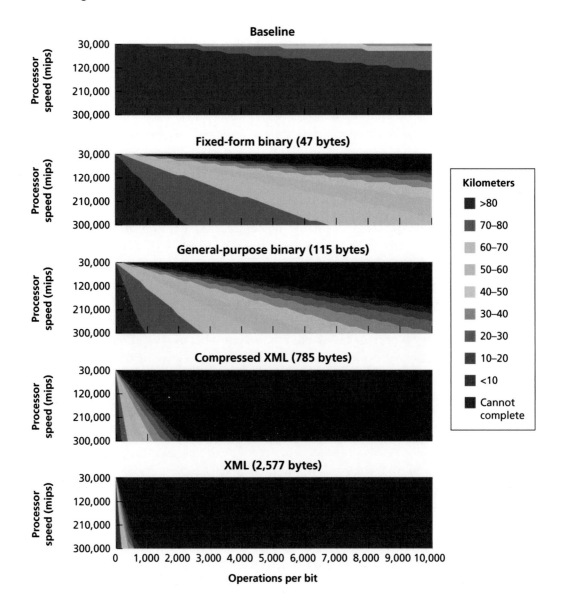

are not a source of delay. The data can move from one subsystem to another in a negligible amount of time. Instead, it is the processing and interpretation of the message once received that causes the delay in initiating the jamming and the resulting potential decrease in effectiveness.

Conclusions

This chapter describes a case of a system with very short reaction times that is optimized for speed. Adding a modest amount of interface overhead will affect operational performance, but the effects could be manageable. However, as the overhead increases, we would expect to see significant operational performance degradation. These types of submillisecond system responses would require a very carefully tailored interface or a much-reduced overhead version of UCCL. Therefore, implementing standard interfaces in these types of systems is higher-risk and requires very careful and detailed engineering analysis to ensure that these operational performance trade-offs are carried out successfully.

   In the next chapter, we look at a vehicle APS. This system requires slower reaction times, in which delays may have a less pronounced effect on mission performance.

CHAPTER SIX

Active Protection System Mission Thread

The MADIS is a family of APSs providing air defense capabilities to protect a ground maneuver force on the move against such threats as unmanned aerial systems (UAS) and fixed-wing and rotary-wing aircraft.[1] MADIS is designed to operate on a pair of tactical vehicles working together: one *Stinger* variant (Mk1) and one *Counter-UAS* variant (Mk2). The Mk1 vehicle is equipped with a turret-launched Stinger missile with EW capabilities, a direct-fire weapon, electro-optical/infrared (EO/IR) sensors, and shoulder-mounted Stinger missiles. The Mk2 is equipped with a a turret-launched counter UAS kinetic weapon, EW capabilities, 360-degree radar, direct-fire weapon, EO/IR sensors, and Beyond Line of Sight gateway/server capability. The Army is developing a similar integrated weapon system, although without linking multiple vehicles.[2]

The Mk1 vehicle interfaces with the Mk2 vehicle via the Adaptive Networking Wideband Waveform.[3] The Mk2 communicates with the Common Aviation Command and Control System to access the common tactical picture (CTP), engagement commands, and weapons alert states. This link is provided via Link 16 running the Joint Range Extension Application Protocol. The CTP includes information about hostile, neutral, and friendly tracks from national, theater, and tactical sensor feeds.

Modeling the Mission Thread and Its Interfaces

MADIS is analyzed here as a generalized APS that affords protection not only to the tactical vehicle on which it is mounted but also to the other vehicle in a MADIS pair,

---

[1]   U.S. Marine Corps, "Ground Based Air Defense," webpage, undated.

[2]   Leonardo DRS Company, "IM-SHORAD," webpage, 2020.

[3]   John Keller, "Harris to Provide Military Special Operations Radios for Sensitive and Covert Missions," *Military and Aerospace Electronics*, January 22, 2014.

as well as to a nearby maneuver force. Coordinating APS capabilities across multiple vehicles may pose networking challenges.[4] The results of this analysis should be viewed as indicative of the family of systems and do not represent the behavior of the actual MADIS itself. They should be used to understand the trade space for these types of systems and not as performance numbers for the specific MADIS. To estimate the impact of the UCCL, we compute the *Minimum Defeat Distance* (MDD)—the smallest launch distance of a threat at which the protection system is expected to succeed in intercepting the threat. The MDD is a commonly used metric to compare APSs.[5] The MDD, measured in meters, is defined as follows:

$$MDD = V \times SRT + IP .$$ 6.1

In Equation 6.1, *V* is the speed of the incoming threat projectile; *SRT* is the *system reaction time*, comprising the time for search, recognition, and identification of the threat until interception by the countermeasure; and *IP* is the *interception point*, the required distance from the exterior of the vehicle to the point of interaction between the threat and countermeasure. In the analysis of APSs, the interception point depends only on the type of hardkill countermeasure.[6] The interception point reflects the fact that safe, kinetic interaction with the threat requires the countermeasure to be some distance away from the protected vehicle when the interaction takes place. In the analyses below, *V* is measured in m/s, *SRT* in seconds (s), and *IP* in meters (m). The interception point for eight APSs reviewed by Haug and Wagner[7] varies from 1 to 30 m.

The MADIS System Interfaces

We evaluated the MADIS APS in two threat scenarios: one in which a cruise missile targets a maneuver force protected by a MADIS pair (Figure 6.1), and one in which a rocket-propelled grenade (RPG) is launched toward the Mk1 vehicle (Figure 6.2). Cruise missile threats appear not to be typically analyzed in the context of APSs.[8]

---

[4]   B. Kempinski and C. Murphy, "Technical Challenges of the U.S. Army's Ground Combat Vehicle Program," in Isak Lundgren, ed., *The Army's Ground Combat Vehicle (GCV) Program*, Hauppauge, N.Y.: Nova Science Publishers, 2013.

[5]   D. Haug and H. J. Wagner, "Active Hard Kill Protection Systems—Analysis and Evaluation of Current System Concepts," *Strategie & Technik*, Autumn 2009.

[6]   Haug and Wagner, 2009.

[7]   Haug and Wagner, 2009.

[8]   They are notably absent from the list of threat reviews in R. Steeb, *Issues for Ground Vehicle Active Protection Systems for the Next Decade*, draft report, U.S. Army Fort Benning Maneuver Center of Excellence, 2017.

Figure 6.1
Cruise Missile Threat Scenario

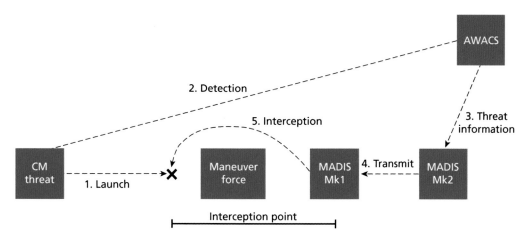

NOTE: AWACS = Airborne Early Warning and Control System.

Figure 6.2
RPG Scenario

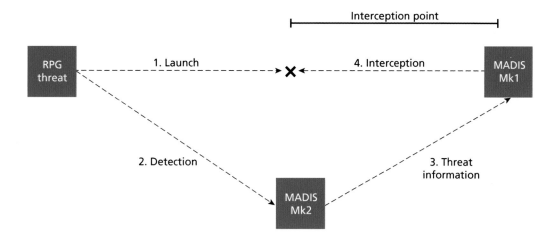

In this case, the interception point is driven by the threat type, rather than by the countermeasure, because interception of a cruise missile payload may result in a wide damage area. For this reason, the interception point assumed in the cruise missile scenario is much larger than in usual APS scenarios (500 m).

In both scenarios, the countermeasure is assumed to be a small missile whose velocity is the same as that of a Stinger missile (750 m/s).[9] This assumption is consistent with the use of missiles in some APSs (e.g., Quick Kill).[10] Neither scenario is explicitly described among GBAD capabilities.[11] The cruise missile and RPG threats might be dealt with by different countermeasures, or the flow of information between tactical vehicles could be different from what is assumed in this analysis.

The Model of the UCCL Interfaces
The flow of information differs between the cruise missile and RPG use cases, resulting in different ways to model *SRT*. Each case is described separately below.

Cruise Missile Threat
Figure 6.1 illustrates the cruise missile threat scenario. The mission vignette starts with a cruise missile being launched toward the Maneuver Force (1). An AWACS detects the incoming cruise missile (2) and relays that information via the CTP to the MADIS Mk2 vehicle (3). The Mk2, in turn, transmits that information to the Mk1 (4), which launches a countermeasure to intercept the cruise missile (5). The notional UCCL runs on a three-node network (AWACS →Mk2→Mk1) in this scenario.

SRT is defined as:

$$SRT = DT + e_a + t_{a \to Mk2} + d_{Mk2} + p_{Mk2} + e_{Mk2} + t_{Mk2 \to Mk1} + d_{Mk1} + p_{Mk1} + c. \quad 6.2$$

---

[9]   See "FIM-92 Stinger MANPADS Man-Portable Surface-to-Air Missile System," fact sheet, Army Recognition.com, November 13, 2020.

[10]   "Hardkill APS Overview," *Below the Turret Ring* blog, January 7, 2017.

[11]   U.S. Marine Corps, undated.

DT is the detection time for a surveillance radar, set to 100 ms.[12] Detection is assumed to be on an airborne platform, given the known difficulty of detecting low-flying missiles from the ground.[13]

e_a is the time for encryption, on the airborne platform, of the message relayed to the Mk2. This time is computed as a function of message size, in bits, and encryption speed. Encryption speed is set to 200 Mbps, based on an existing standard.[14]

$t_{a \rightarrow Mk2}$ is the time to transmit the message from the airborne platform to the Mk2 via Link 16. This time is a function of message size and bandwidth of Link 16, set here to 115.2 kbps.[15]

d_{Mk2} is the time needed to decrypt the message received on the Mk2. This time depends on message size and decryption speed, set again to 200 Mbps.

p_{Mk2} is the time to process the incoming data from the airborne platform on the Mk2 vehicle. We make no specific assumptions about the nature of the processing but note that it would include at least tracking an incoming threat and determining whether it will pass through a protected area.[16] Processing time is computed by multiplying the message size in bits with an estimate of the number of operations per bit and dividing the result by the processor speed (in mips). The number of operations per bit depends on the kind of processing performed on the data. For example, a previous report estimates the number of operations per bit to process internet headers to be at least 100, and potentially much larger.[17] We treat this estimate as a lower bound and, instead, set the number of operations per bit to 100,000 as an upper bound, for processing either headers or data. The mips is set to 49,360, corresponding to an Intel Core i7 7500U.[18] As a worst-case analysis, it is assumed that the Mk2 vehicle needs to

---

[12] Simon Kingsley and Shaun Quegan, *Understanding Radar Systems*, Vol. 2, Chennai, India: SciTech Publishing, 1999.

[13] Defense Industry Daily staff, "JLENS: Co-ordinating Cruise Missile Defense—and More," *Defense Industry Daily*, February 13, 2017; Lee O. Upton and Lewis A. Thurman, "Radars for the Detection and Tracking of Cruise Missiles," Lincoln Laboratory Journal, Vol. 12, No. 2, 2000.

[14] General Dynamics Mission Systems, "TACLANE-Micro (KG-175D) Encryptor," fact sheet, 2020.

[15] Viasat, "MIDS-LVT Terminals," webpage, 2021.

[16] Viasat, 2021.

[17] Tucker and Hinton, 2011, p. 828. In this reference, the number of operations per bit for internet packer header recognition is reported to be on the order of 100, and on the order of 10,000 and higher for a single signal processing application. We set it to 100,000 given that multiple data and signal processing algorithms are required in our system.

[18] "Instructions per Second," *Wikipedia* entry, webpage, updated February 22, 2021.

decrypt the message from the AWACS and encrypt it again to send it to the Mk1. The time for encryption on the Mk2, e_{Mk2}, is set to the same value as e_a.

$t_{Mk2 \rightarrow Mk1}$ is the time to transmit the message from the Mk2 to the Mk1 via the Adaptive Networking Wideband Waveform link. This time is a function of message size and radio bandwidth, set to 500 kbps.[19]

d_{Mk1} and p_{Mk1} are the times to decrypt and process the message on the Mk1 and are set equal to d_{Mk2} and p_{Mk2}, respectively.

c is the time for the countermeasure from the Mk1 to reach the interception point. We assume a worst-case scenario where the Mk1, the maneuver force, and the cruise missile launch point form a single line, with the maneuver force located somewhere in the middle. Compared with other geometrical placements, this entails the longest possible interception path for the countermeasure. In this setting, the countermeasure needs to traverse the distance between the Mk1 and the maneuver force and then to the interception point. Without further constraining information, the distance from the Mk1 and the interception point is assumed to be 500 m. The time to cover the distance is computed as the ratio of that distance to the speed of the countermeasure.

Rocket-Propelled Grenade Threat

Figure 6.2 illustrates the RPG scenario. The mission vignette starts with an RPG being launched toward the Mk1 (1). The Mk2 vehicle detects the incoming RPG (2) and informs the Mk1 vehicle (3). The Mk1 vehicle then launches a countermeasure to intercept the RPG (4). The notional UCCL runs on a two-node network (Mk2→Mk1) in this scenario.

The calculation of *SRT* is considerably simplified, relative to that for the cruise missile threat:

$$SRT = DT + e_{Mk2} + t_{Mk2 \rightarrow Mk1} + d_{Mk1} + p_{Mk1} + c \ . \qquad 6.3$$

Here, time for *DT* is set to 100 ms, which is on the low end—although not the lowest—of *SRT*s for various APSs reviewed in Haug and Wagner (2009). All other quantities are determined as in the cruise missile case, except for c. The interception point is set to 30 m, consistent with typical values used for RPG threats. We assume the same countermeasure as in the cruise missile case—a projectile with the velocity of

---

[19]   Mark Turner and Ken Dingman, "Developing SCA Based Wideband Networking Waveforms," presentation slides, Harris Corporation, 2011.

a Stinger missile. The Quick Kill system, for example, uses guided or unguided missiles as countermeasures for RPGs.[20]

Equation 6.3 and the geometry displayed in Figure 6.2 are consistent with two RPG scenarios:

1. The RPG is targeted at the Mk1 vehicle, but that vehicle's sensors are disabled. In this case, the Mk2 vehicle provides auxiliary sensing capabilities to the Mk1.
2. The RPG is targeted at the Mk2 vehicle and is detected by that vehicle's sensors, but the vehicle is out of countermeasures to intercept the threat. In this case, the Mk2 vehicle provides targeting coordinates to the Mk1 vehicle and instructs it to intercept the threat. This could happen, for example, if multiple RPGs were simultaneously fired at the Mk2 and the vehicle had already launched all loaded countermeasures. In this case, the Mk1 provides the Mk2 with an improved capability to deal with salvos.

The second scenario—the ability to respond quickly to multiple incoming threats—is probably most relevant to APS systems, because long reload times significantly increase the MDD.[21]

Modeling Assumptions

The MDD is defined as the distance between the threat's launch point and the targeted force. In the cruise missile scenario, this is the distance from the maneuver force to the point of launch of the cruise missile. In the case of the RPG, it is the distance from the Mk1 vehicle to the point of launch of the RPG.

The speed of the threat (V in Equation 6.1) is defined using information available in the literature. Table 6.1 lists three examples each for cruise missile and RPG threats. Threats highlighted in gray—Kh-59ME for cruise missiles and RPG-29 for RPGs— were used in the current numerical studies. Only the reported speeds are used in our calculations. The maximum operational ranges are used to check that the resulting MDDs are within reasonable use ranges.

The MDD is computed with Equations 6.1 or 6.2, for increasing message sizes and number of operations per bit. The message size corresponds to the sum of the sizes of the data and of the header. The size of the headers is varied from 0 to 256,000 bits. The size of the data is set to 47 bytes, which corresponds to the mean size of FFB mes-

---

[20] "Hardkill APS Overview," 2017.

[21] The reload time of 1.5 seconds for the Trophy system is estimated to increase the MDD to 450m in the case of RPG-7 threats. See "Hardkill APS Overview," 2017.

Table 6.1
Threat Types

Threat Class	Threat	Speed (m/s)	Range	Source
CM	Hermes	1,300	100km	Deagle.com, homepage, undated-a.
	Hongniao-1 (H1-A)	274	600km	Missile Defense Project, "Hong Niao Series (HN-1/-2/-3)," *Missile Threat*, database, Center for Strategic and International Studies, August 12, 2016, last modified November 26, 2019
	Kh-59ME[a]	292	115km	Deagle, undated-a
RPG	RPG-7 with PG7VR	100	200m	Maxim Popenker, "Modern Firearms," webpage, undated (range); Haug and Wagner, 2011 (velocities)
	RPG-7 basis type	200	200–500m	
	RPG-29	450	500m	

NOTE: CM = cruise missile.

[a] The Kh-59ME is an export variant of the Kh-59 that can destroy both ground and surface ship targets (GlobalSecurity, "Raduga Kh-59 (AS-13 Kingbolt) and Kh-59M (AS-18 Kazoo)," webpage, updated October 18, 2015), in contrast to the Kh-59MK (Deagle, "KH-59," webpage, undated-b), which is designed to destroy exclusively ships.

sages reported elsewhere in the context of the analysis of typical implementations of message formats.[22]

For reference, the smallest useful message would need to convey at least the instant position (latitude, longitude, altitude) and velocity of the threat along each spatial dimension, for a total of six coefficients. If each coefficient were encoded as a Double Double (128 bits), the total data size would be 96 bytes.

Finally, we make the following simplifying assumptions on the geometry of the scenario:

1. The missile turret on the Mk1 already points toward the threat (i.e., the turret does not need to be rotated to adjust either azimuth or elevation).
2. For the RPG scenario, there are no occluding obstacles between the threat and the MADIS system that might impair detection by the Mk2 vehicle.
3. The threats follow a straight-line trajectory to their target and move at a constant, maximum velocity.

---

[22] Rutledge et al., forthcoming.

Table 6.2 summarizes the parameters used in modeling.

Modeling Results/Critical Step Analysis

Since the APS sequence is composed of very few rapid steps and all must be accomplished in a specific order for successful interception, this analysis encompasses the entire sequence, and no separate critical-step analysis was done. In the context of this mission thread, each step is a critical step.

Keeping data size constant and varying header size and number of operations per bit, we computed the MDD results in the surfaces depicted in Figures 6.3 and 6.5 for cruise missile and RPG threats, respectively. The message sizes corresponding to the four data formats in Table 6.3 are overlaid as horizontal cross-sections on the surfaces. These should not be interpreted as estimates of overhead for each implemen-

Table 6.2
Modeling Parameters

Parameter	Units	Value
DT	ms	100
Encryption speed	Mbps	200
Decryption speed	Mbps	200
Link 16 bandwidth	Kbps	115.2
ANW2 bandwidth	Kbps	500
Processor speed	mips	49,360
Number of operations per bit to process data and header	operation	0 to 100,000
Interception point (RPG)	m	30
Interception point (CM)	m	500
Threat velocity (RPG)	m/s	450
Threat velocity (CM)	m/s	292
Countermeasure velocity	m/s	750
Header size	bits	0 to 256,000

NOTE: CM = cruise missile; ANW2 = Adaptive Networking Wideband Waveform.

Figure 6.3
MDD as a Function of Operations per Bit and Header Size for Cruise Missile Threats

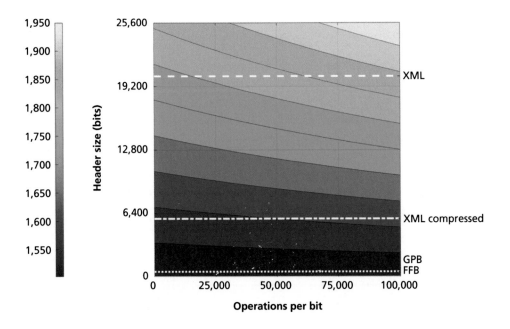

Table 6.3
Four Data Formats

Data Format	Message Size (Bytes)
XML	2,577
Compressed XML	785
General-purpose binary (GPB)	115
FFB	47

tation approach, but as a range of possible input values. APS implementations of each approach will differ and depend on several design decisions that are beyond the scope of this study. Hence, when we reference performance of an XML implementation as compared with an FFB, we are referring to these representative values of message sizes and not to actual XML or FFB implementations of a standard representing real APS system message sizes.

Figures 6.4 and 6.6 illustrate the MDD broken down into "phases of engagement" for cruise missile and RPG threats, respectively, for the four data formats. The highest number of operations per bit is assumed (i.e., the data shown correspond to the MDD measured at the right side of each horizontal cross-section). The breakdown is across the six phases defined in Table 6.4.

Cruise Missile Threats

Figure 6.3 shows the MDD surface for cruise missile threats according to the Kh-59ME velocity profile. Even when header size and number of operations per bit are at their highest, the resulting MDD (~2,000m) is well within the operational range of this kind of threat (115km). The same figure shows the XML, XML compressed, GPB, and FFB values for message sizes as horizontal lines.

Figure 6.4 displays the components of the MDD for the four data formats, assuming the highest number of operations per bit. A GPB value has a barely notice-

Figure 6.4
MDD Breakdown Across Phases of Engagement for Cruise Missile Threats

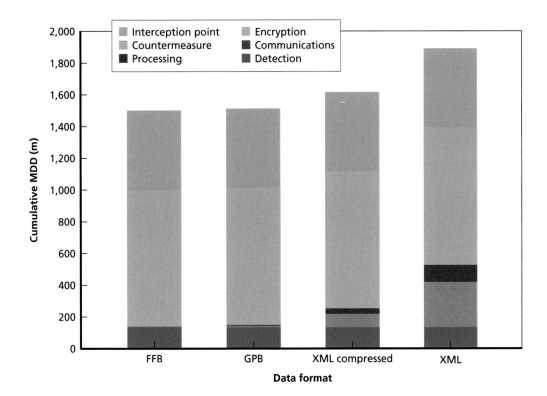

Table 6.4
Definitions of the Phases of the MDD

Phase	Definition (CM)	Definition (RPG)
Detection	VDT	VDT
Communications	$V(t_{a \rightarrow Mk2} + t_{Mk2 \rightarrow Mk1})$	
Encryption	$V(e_a + d_{Mk2} + e_{Mk2} + d_{Mk1})$	$V(e_{Mk2} + d_{Mk1})$
Processing	$V(p_{Mk2} + p_{Mk1})$	Vp_{Mk1}
Countermeasure	Vc	Vc
Interception point	IP	IP

Figure 6.5
MDD as a Function of Operations per Bit and Header Size for RPG Threats

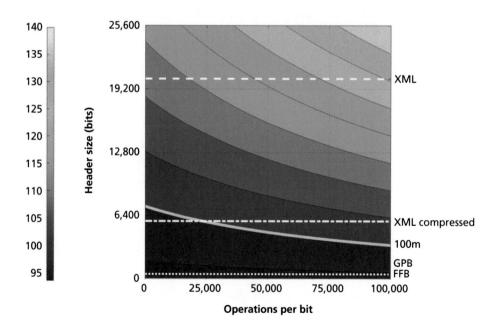

Figure 6.6
MDD Breakdown Across Phases of Engagement for RPG Threats

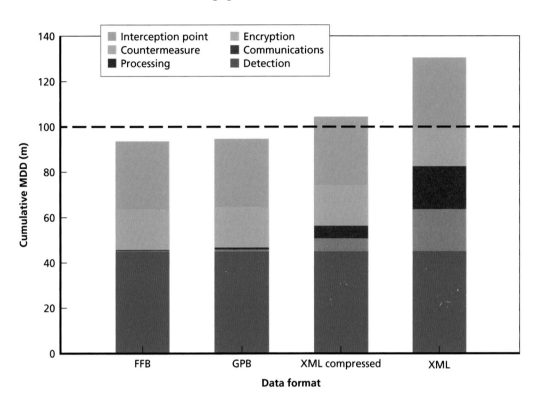

able impact on the MDD relative to the FFB value. The Standard XML value, in particular, incurs a large penalty due essentially to communications latency, resulting in an increase of nearly 400m in the MDD relative to the FFB value.

RPG Threats

Figure 6.5 shows the MDD surface for RPG threats based on the RPG-29 velocity profile. Even when header size and number of operations per bit are at their highest, the resulting MDD (~140 m) is well within the operational range of this kind of threat (500 m).

Figure 6.6 displays the components of the MDD for the four data formats, assuming the highest number of operations per bit. As for the cruise missile case, the impact of the GPB value is barely noticeable on the MDD relative to an FFB value. A Standard XML value incurs a large penalty, increasing MDD by nearly 40m relative to FFB for that message size. The relative contribution of communications delay to the

MDD is smaller than in the cruise missile case because of the added hop over Link 16 in the latter case.

RPGs are a typical focus of APS systems;[23] hence, it is possible to comment on the suitability of the MDDs obtained in this case study in the light of publicly available requirements. APS tests were conducted at the Naval Surface Warfare Center, Dahlgren Division in March 2006 with threats launched at 100m.[24] A threat range of 100m was also assumed in a previous study on survivability in urban operations.[25] Assuming 100m as a requirement, we show in Figure 6.6 that, at the highest number of operations per bit considered, only Standard XML and compressed XML values fail to remain below the required MDD. Figure 6.6 shows that compressed XML might fall below the 100m requirement for low-enough number of operations per bit.

Another comparison that can be drawn is with the MDD of the Quick Kill system, another APS that relies on missile countermeasures.[26] The MDD for Quick Kill was previously calculated at 187.5m for RPG-29 threats, which exceeds all MDDs reported in Figure 6.5.[27] Hence, despite added latency caused by the network, the MADIS configuration in this scenario would still be preferable to having each vehicle carry an independent Quick Kill system with these performance parameters.

Conclusions

APS performance is driven by threat, operational context, and the performance of the component systems—not by additional overhead of a UCCL. For stressing cases of higher-end threats, and for very tight MDD performance requirements, one should do a careful, detailed engineering analysis to ensure that these operational performance trade-offs are carried out successfully. For example, we see that neither Standard XML nor compressed XML satisfy a 100-m MDD requirement for high-speed RPG

---

[23]   Andrew Feickert, "Army and Marine Corps Active Protection System (APS) Efforts," Washington D.C., August 30, 2016.

[24]   Committee on Armed Services, Tactical Air and Land Forces Subcommittee, *Combat Vehicle Active Protection Systems*, in U.S. House of Representatives, 109th Congress, second session, Washington, D.C.: U.S. Government Publications Office, September 1, 2006.

[25]   L. Wong, "Systems Engineering Approach to Ground Combat Vehicle Survivability in Urban Operations," thesis, Naval Postgraduate School, Monterey, Calif., 2016.

[26]   Peter Ramjug, "Raytheon's Quick Kill Active Protection System Defeats One of the Most Lethal Armor-Piercing Rocket Propelled Grenades," press release, Raytheon Co., January 9, 2013.

[27]   Haug and Wagner, 2009.

threats under the assumptions modeled. But a GPB interface seems to have minimal impact on operational performance, even under a wide set of assumptions. No clear MDD requirement is available for cruise missile threats, but the analysis shows that a UCCL—even if poorly defined—would not exert a driving influence on the MDD. In most cases, interface inefficiencies are not likely to be the main contributor to the MDD, even assuming unrealistically high overhead sizes. A carefully designed UCCL should be able to meet operational performance needs under a wide range of cases.

CHAPTER SEVEN

Ballistic Missile Defense Mission Thread

Missiles designed to fly a ballistic trajectory threaten territory far from their launch sites and are hard to defeat but have key periods of vulnerability that a layered defense system exploits.[1] All ballistic trajectories have an initial period of powered flight (the boost phase), at the end of which the missile orients and releases one or more reentry vehicles (RVs), decoys, and other debris into a parabolic uncontrolled flight determined primarily by the force of gravity (the midcourse phase) until they reenter the Earth's atmosphere and encounter the dynamic forces of wind and atmospheric drag (the terminal phase). While it may be desirable to intercept a ballistic missile during the boost phase while it is still over adversary territory, powered flight is quite short (typically between 1 and 6 minutes, depending on the missile), leaving little time to detect and establish a good estimate (called a *track*) of its future position. Intercepting an RV during the midcourse state is made simpler by the lack of forces perturbing the trajectory (the forces of gravity being relatively well known), but it is often difficult to distinguish the RV and its associated warheads from decoys and debris because they all tend to move similarly under the slight forces of gravitational pull. Also, depending on range to the target, the missile may not leave the atmosphere or may leave it only briefly, meaning that not all missile attacks offer the opportunity for midcourse intercept. Intercepts during the terminal phase provide more opportunities to discriminate decoys and debris from the actual warhead(s), but also more disturbance forces to complicate the task of establishing accurate track estimates. The National Academy of Sciences notes that the discrimination task becomes easier between 70 and 100 km above sea level, but that the track estimation task becomes much harder below 40 km.[2]

---

[1] Kenneth Werrell provides a historical overview of the challenges of BMD in "Hitting a Bullet with a Bullet: A History of Ballistic Missile Defense," Airpower Research Institute Research Paper 2000-02, Air University College of Aerospace Doctrine, Research and Education, Maxwell Air Force Base, Ala., 2000.

[2] National Research Council, *Making Sense of Ballistic Missile Defense: An Assessment of Concepts and Systems for U.S. Boost-Phase Missile Defense in Comparison to Other Alternatives*, Washington, D.C.: National Academies

Of course, all of the above becomes much harder if we need to defend against multiple incoming attacks.

Figure 7.1 illustrates the overall approach used for BMD in the United States today. The first step is to detect and classify the missile threat, typically the task of long-range radars and satellite-based infrared detectors.[3] Each sensor that detects a missile launch transmits measurement information to a C2 battle management center (C2BMC), where it is combined with other information (such as the local gravity field, winds aloft, threat unique parameters, etc.) to predict the forward trajectory and cue downstream sensors closer to the predicted destination. The cued sensors initiate a search for the missile using the track predictors and, if it is found, begin to transmit their track data to the C2BMC. These additional track data from different geometries greatly improve the accuracy of the predicted trajectory. After the RV is released, the

Figure 7.1
The BMD Kill Chain

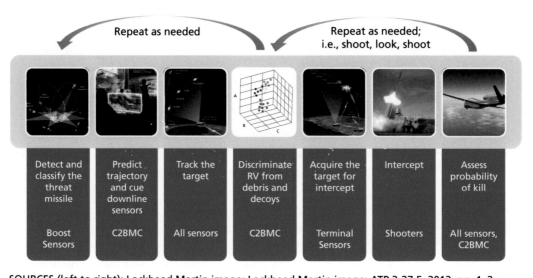

| Repeat as needed | | | | Repeat as needed; i.e., shoot, look, shoot | | |

Detect and classify the threat missile	Predict trajectory and cue downline sensors	Track the target	Discriminate RV from debris and decoys	Acquire the target for intercept	Intercept	Assess probability of kill
Boost Sensors	C2BMC	All sensors	C2BMC	Terminal Sensors	Shooters	All sensors, C2BMC

SOURCES (left to right): Lockheed Martin image; Lockheed Martin image; ATP 3-27.5, 2012, pp. 1–2; RAND; ATP 3-27.5, 2012, pp. 1–2; Ralph Scott, Missile Defense Agency; Air Force photo. The steps of the kill chain are derived from David Gompert and Jeffrey A. Isaacson, *Planning a Ballistic Missile Defense System of Systems: An Adaptive Strategy*, Santa Monica, Calif.: RAND Corporation, IP-181, 1999.

Press, 2012.

[3] Shorter-range radars may be of use in the boost phase if launch sites are near borders or if the radars are ship-mounted.

sensor network provides data on each object in the threat cloud, and the C2BMC can begin to run algorithms to try to distinguish the warheads from decoys and debris.

Terminal sensors, which for theater defense are often radars co-located with an interceptor such as the U.S. Terminal High Altitude Area Defense (THAAD) or Aegis systems, are cued to acquire the target for intercept. Upon authorization to shoot, the interceptor is launched toward the target. After a target is intercepted, a variety of sensors are used to gather information to assist in assessing the probability of kill (P_{kill}). This enables a shoot, look, shoot operations concept if time allows.

When considering how a UCCL could impact the BMD mission, we need to address the three underlying factors that impact the degree of difficulty of that mission. The first is the overall timeline predicated on the range to target. The timeline to detect and defend against a short-range attack—for example, from North to South Korea—is much shorter than the timeline to detect and defend against a long-range missile attack on the U.S. homeland.[4] The second factor is whether the missile is launched from a known fixed site that has been precisely characterized in terms of position and local gravity field versus from a mobile launcher whose initial position must be determined from sensor data. The last factor is the quantities of warheads and decoys (or other penetration aids) deployed by the missile. As these increase in number, it becomes harder to find the lethal warheads amid the threat cloud.[5]

In considering how a UCCL could affect the BMD mission, we defined a stressing case to consider:

- tactical BMD traveling at 5 km/second for 3,500 km, assumed boost phase of 1 minute with a total travel time of ~11 minutes
- launched from a mobile or previously unknown site
- sophisticated penetration aids that cannot be discriminated until the kill vehicle is 25 m from the target and the target is less than 70 km above sea level.[6]

---

[4] An ICBM travels roughly 7 km/second and covers a 10,000-km range (23 minutes of flight time). Tactical ballistic missiles are generally defined as those traveling less than 5 km/second over a 3,500-km range (11 minutes of flight time). Werrell, 2000, p. 60.

[5] Dimensions of threat are derived from Gompert and Isaacson, 1999.

[6] Werrell states that the Iraqi Scud missiles used in the Gulf War "made good use of high-fidelity East German decoys, which reportedly could not be distinguished from the real items at distances greater than 25 yards." While discrimination has undoubtedly improved in the intervening years, we hypothesize that so have decoys and so used the 25 meters in defining our stressing case. Werrell, 2000, p. 36.

Modeling the BMD Mission Thread

Having defined a stressing case, the next step in understanding how a UCCL could affect the BMD mission thread is to build a model of the mission interfaces that allows us to vary characteristics of interest to our research. To fully realize the advantages of an any sensor, any shooter architecture for tactical BMD may require a UCCL that facilitates ad hoc or preplanned integration of sensors and shooters on the battlefield. In that case, complex protocols are required to maintain synchronization and smooth reliable interactions between systems. The following list describes a few of those potential protocols, each of which adds messages and complexity to the interfaces of an SoS:

- If the set of systems that will interact is not fully predetermined, a *discovery protocol* allows new systems to be discovered and join into the SoS. The more diverse the set of systems allowed to join, the more complex this discovery protocol will be.
- For BMD, clock synchronization across the SoS is critical if tracks are to be interpreted correctly and if discrimination algorithms are to correctly identify the warhead among decoys and debris. Therefore, any UCCL that supports the BMD mission thread will require a high-accuracy time synchronization protocol.
- The SoS will also need a means to agree upon a geospatial reference frame, often implemented by a *reference frame synchronization protocol*.
- A QoS protocol might also be needed, as we saw in the case of DDS in Chapter Three.

All of the inefficiencies described above will require additional data to flow between systems and to be processed by each system. We model these inefficiencies using three variables:

1. header overhead B_h, in bits. This is the added information that enables addressing each element of the SoS—which we will call a *node*—and identifying it by mission and/or capabilities.
2. additional data to be exchanged, $Data_{m,k}$. This term accounts for additional data (e.g., larger common data fields or extra synchronization protocols) exchanged between any two nodes m and k.
3. Additional data transformations, $DataTR_{m,k}$. This term accounts for additional processing time needed to transform data between coordinate systems, units of measure, or formats.

A mission thread is modeled as a directed graph $D = (V,E)$ with nodes u_k and edges $\{(u_m, u_k)\}$ that equal 0 if two nodes do not have a communications path between them and 1 if a path exists. Each node is either a sensor node, a shooter node, or a C2 node. We model the communication between any two connected nodes m and k as a path with a data rate, $DR_{m,k}$, in bits per second. At each node k, we model a CPU and an encryption/decryption device.[7] While all extra data must be processed by the CPU (and optionally by the encryption/decryption device), not all processing is the same. For example, processing headers is less computationally expensive than signal processing algorithms. Hence, we use separate complexity factors for processing headers (HCF_k) versus more general data (DCF_k). The number of transforms that need to be executed on the additional data is modeled by N_T, and the complexity of processing those transforms is modeled as TCF_k.[8] Total additional delay from implementation of a standard interface is then a function of (1) extra CPU cycles to process and/or transform the data, (2) extra time needed for encryption/decryption processing cycles, and (3) extra time needed to transmit the data on the wire. The equations for these factors are given below:

- The extra CPU cycles to process and/or transform the data at node k are given by the equation,

$$CPUCycles_k = B_h \times HCF_k + Data_{m,k} \times DCF_k + N_T \times DataTR_{m,k} \times TCF_k.$$

- The processing delay is the total extra CPU cycles divided by the processor speed:

$$\frac{CPUCycles_k}{CPU_k} .$$

- The encryption/decryption delay is the amount of data to be sent over the channel, divided by the processing rate of the encryption. Note that the data to be transformed are processed locally (i.e., they are not sent over the channel) and thus are not included in the calculation.

---

[7]   The processing speed of each CPU is given as $CPUCycles_k$ and the encryption/decryption speed at each node is modeled as $CryptoRate_{m,k}$ (both are specified in bits processed per second).

[8]   *Complexity factors* represents, on average, how many additional CPU cycles are required to process additional data. All complexity factors are specified in CPU cycles per bit.

$$\frac{\left(B_h + Data_{m,k}\right)}{CryptoRate_{m,k}} \; .$$

- The data transmission delay is also the amount of data to be sent over the channel but now divided by the data rate in bits per second:

$$\frac{B_h + Data_{m,k}}{DR_{m,k}} \; .$$

- The processing delay is the total number of operations performed on all the data divided by the processor speed in operations per second. The total delay between nodes m and k is the summation of the above delays.

$$Delay_{m,k} = \left[\left(\frac{B_h + Data_{m,k}}{DR_{m,k}}\right) + \left(\frac{CPUCycles_k}{CPU_k}\right) + \frac{\left(B_h + Data_{m,k}\right)}{CryptoRate_{m,k}}\right]$$

- Finally, we sum all steps in the mission thread to compute total delay:

$$Delay = \sum_{m,k=1}^{n} Delay_{m,k} \; .$$

These equations were implemented using the Python package NetworkX. The model ingests information about each node and edge in the directed graph, as well as the size of each message that will be sent. Other inputs to the model include the CPU speed, the effective bandwidth between nodes, the encryption rate, and complexity factors. It then reads in a starting location and a set of ordered destinations that the message must pass through sequentially. The model determines the shortest path to the next destination and records the amount of time required to transfer and process the message. We repeat this for every destination in the set, with the new start location being the previous destination. Finally, it produces graphs depicting the time required for each combination of messages to be passed to the final destination node over the range of input parameters. The sequence of messages passed from system to system is based on the mission threads described in Figures 7.7–7.9 later in this chapter.

Limitations of Our Model

The BMD mission threads we modeled are extracted from publicly available information, with the intent of understanding how different UCCL implementations might affect performance of the SoS. We did not attempt to estimate system parameters or create broad estimates of mission performance. For example, we did not attempt to estimate the data rate of each link, or the processing power of each CPU in a representative BMD deployment of systems. Instead, we assigned one CPU processing rate and one data rate to all nodes and node pairs, respectively, and varied them *all together over a range of possible processing and data rates*. In cases where we kept these two parameters constant, we used an average estimate and varied other parameters to highlight the dependencies between parameters. The intent was to estimate *performance sensitivity* to the parameters of interest, not to estimate the actual performance of any given BMD SoS. Any judgments offered relate to *potential areas of performance risks* and should not be interpreted as a validation of different designs with respect to mission performance requirements.[9]

To root our study in the real world, we represented the range of ways a standard interface could be implemented (presented in Chapter Four), using four representative message sizes measured by MIT Lincoln Labs in Army field tests.[10] These were an XML implementation of 2,577 bytes per packet, a compressed XML of 785 bytes per packet, a GPB based on Google proto buffers of 115 bytes per packet, and an FFB of 47 bytes per packet. These should be interpreted not as estimates of overhead for each implementation approach but simply as point values that define a range of implementation options. Actual BMD implementations of each approach would differ and depend on design decisions beyond the scope of this research. Hence, when we reference performance of an XML implementation as compared with an FFB, we are referring to these representative message sizes and not to actual XML or FFB implementations of a standard.

Nominal Model Inputs
When not varying a parameter, we set it to a nominal value. The nominal complexity factor for header and data processing was set to 100 operations per bit. This is a typical value for processing IP headers that would be needed to pass relatively simple data, such as a target track.[11] Note that complexity factors can vary to more than 10,000 operations per bit for signal processing algorithms,[12] but we assume here that we are

---

[9] While a model such as ours could be used to obtain estimates of actual system performance, doing so would require detailed (and likely classified) system parameters, limiting the distribution of our results.

[10] DeWeck, 2015.

[11] Tucker and Hinton, 2011.

[12] Tucker and Hinton, 2011.

exchanging processed target tracks (i.e., we are assuming raw signal data are processed locally). The nominal value for CPU speed was set to 10,000 mips, roughly comparable to a mid-2000s processor. The nominal data rate between nodes was set to 1 Mbps, typical for a DoD satellite communications system. We set the encryption/decryption rate at 20 Mbps, typical of a lower-end tactical device, such as the KG-175X.

As discussed previously, we used four representative message sizes derived from experimental data. For each of these sizes, we varied the other modeling parameters to estimate how widely delay varies as a function of each parameter and each packet size. Finally, we made a set of runs with the data rate between nodes set to the minimum observed in this system—the satellite link(s) that transmits data from a Space-Based Infrared System (SBIRS) to a ground-based C2 node (Table 7.1). In a full nuclear scintillation environment, data transmissions on these links may be as low as 75 bps.

Sensitivity to CPU Speed and Computational Complexity
BMD algorithms are often quite complex (such as those used for track estimation and target discrimination) and may require a high number of operations per bit of data processed. To understand the sensitivity of the different interface implementations to computational complexity, we kept all other parameters at their nominal values and varied the complexity factors (i.e., operations per bit required to process the message) over the range from 100 to 10,000 operations per bit. As shown in Figure 7.2, for implementations that result in a smaller packet size, total delays are relatively insensi-

Table 7.1
Modeling Parameters

Parameter	Units	Value
Encryption speed	Mbps	20
Decryption speed	Mbps	20
Nominal link data rate	Mbps	1
Range of link data rate	bps	75 to 2,000
Nominal processor speed	mips	10,000
Range of processor speed	mips	1 to 20
Nominal complexity factor	Operation per bit	100
Range of complexity factor	bits	100 to 15,000

tive to computational complexity—overall mission delay stays under a fraction of a second. Even for the worst-case implementation we modeled (human-readable XML), the total delay varies on the order of a few seconds. For a mission thread like BMD that typically takes minutes to complete, these additional delays appear relatively inconsequential. We caution, however, that while we found little sensitivity of total mission delay to computational complexity over the range of interface implementations modeled, if a specific step in the BMD process has especially complex computations and the CPU is operating at near capacity, then the implementation of an interface standard may indeed adversely affect the mission by requiring additional header processing and, thus, overwhelming the processor.

BMD systems often include legacy systems with older, lower-speed CPUs. In some cases, even modern high-speed CPUs can be overwhelmed by other computations that are part of the mission execution, leaving little CPU space available for message processing. To examine the effect of CPU speed on total mission delay, we set all other parameters to their nominal values and varied the CPU for 1 to 20 mips, which corresponds to a late 1970s to late 1980s CPU processing power, or CPUs that are over-

Figure 7.2
Mission Thread Delay Versus Computational Complexity Factors

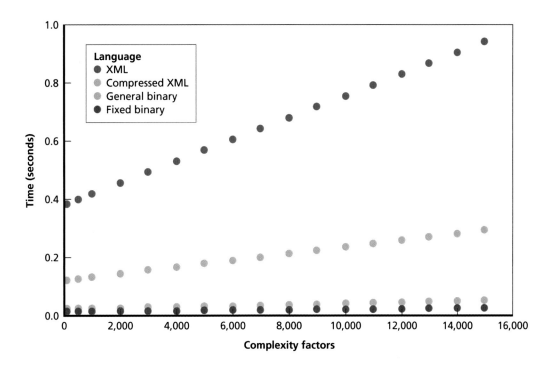

burdened. For the smaller packet sizes, total delays are relatively insensitive to CPU speed, and overall mission delay stays within seconds for all cases run (Figure 7.3). But for larger packet sizes, such as those we might expect from an XML implementation, delays are significant at very low CPU speeds. Therefore, we find that BMD mission performance may be adversely affected if slow or overburdened CPUs are used with an inefficient interface implementation. Additional study is warranted.

Sensitivity to Data Rate and Data Packet Sizes

As discussed earlier, the link from SBIRS to its ground station may drop as low as 75 bits per second in the case of a nuclear detonation causing atmospheric scintillation. Given that this appears to be a worst case, we examined the additional delay to complete this single step of the mission thread as a function of the message sizes that we might expect from different interface implementations. Results are shown in Figure 7.4 for each interface implementation with a varying link data rate between 75 and 2,000 bps. Under these conditions, message size has a significant impact on delay as the data rate of the link decreases. The information from SBIRS alerts the ground C2

Figure 7.3
Mission Thread Delay Versus CPU Speed

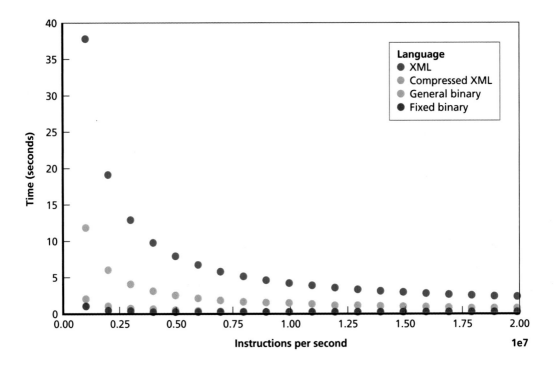

node that a missile launch has been detected and occurs during the boost phase. Typically, this boost phase lasts between 1 and 6 minutes. At the lower bound, an additional delay of 30 seconds could be significant, in that it could delay notification to downstream sensors that operate during the midcourse phase. At the upper bound, an additional 30-second delay in receipt of data from SBIRS is unlikely to affect execution of the mission. However, bandwidth-constrained links in other steps of the BMD mission might be more sensitive to delay. In the critical step analysis section of this chapter, we look at another bandwidth-constrained link—that from the radar to the interceptor—and discuss the impact an inefficient messaging implementation might have on it.

Building on these sensitivity analyses, we focused our attention in the critical step analysis section on steps within the BMD mission thread that include (a) bandwidth-constrained links and/or (b) complex data processing executing on constrained CPUs.

Figure 7.4
Single Step Delay, in Seconds, as a Function of Link Data Rate and Interface Implementation

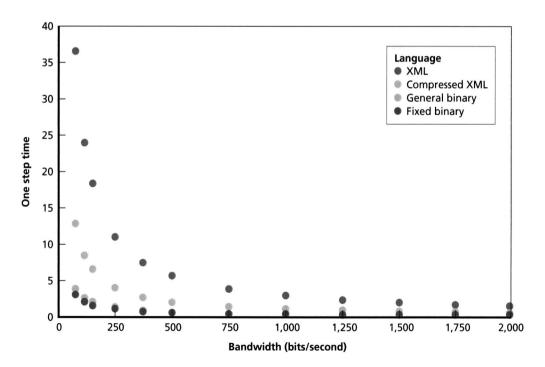

Worst-Case Scenario for Impact of Latency on Medium-Range Ballistic Missile Tracking

In the previous section, we explored the sensitivity of delays to different parameters as a function of the size of message that results from some standard interface implementations. In this section, we assess the sensitivity of the BMD mission to latency (i.e., delay) in the transmission of data needed to form a track estimate for a typical threat or a delay in the transmission of the track estimate to forward elements in the mission thread. If unmitigated, any latency in the collection and use of these data—including the latency added by a UCCL—will have a direct and undesirable impact on track quality and, ultimately, the outcome of a missile intercept. The impact of measurement (or track) latency on track quality is a complex function of sensor capabilities and geometry, as well as concept of operations and the phase of missile flight.

Since latency is a known problem in BMD systems, the typical BMD system is designed to mitigate its effect. Sensor(s) continually share updated measurement and track data at a specified update rate. These data arrive at their intended destination, albeit with some delay. It is vital, however, that the receiving algorithms have unambiguous and accurate knowledge of the time at which the data were valid. Therefore, each critical data item includes a time stamp to enable the receiving algorithms to properly understand when to apply new information to correct the estimated state space of the target (this is often referred to as the *trajectory* of the target). If sensor messages become so large, however, that they stress the network bandwidth and result in very stale or lost messages, the result is effectively a *sensor blackout*; i.e., the tracking algorithms receive no measurements, in which case they continue to propagate the last known good data to estimate the state of the target. The longer the estimates are propagated without being corrected by new measurements or tracks, the more inaccurate the estimated tracks become. We use this sensor blackout scenario to assess the worst-case effect that latency can have on track quality (e.g., position error). Where possible, we identify latency *boundaries* or *conditions* that (1) do not affect the system and (2) require further and more detailed analysis to more accurately assess the impact of latency.

To apply this analysis, consider the tracking of a medium-range ballistic missile (MRBM). A simplified flight profile of a DF-21 MRBM is considered in Figure 7.5. The range of the missile is 1,700 km, with an apogee of 500 km and maximum velocity of ~Mach 12. In the nominal case, shown in the figure, no sensor blackout occurs, and the track algorithms regularly update the estimated position and velocity of the missile using newly received data to approximate the true position and velocity. Downstream sensors or the interceptor itself can be *cued*; that is, they propagate the

Figure 7.5
Flight Profile of an MRBM

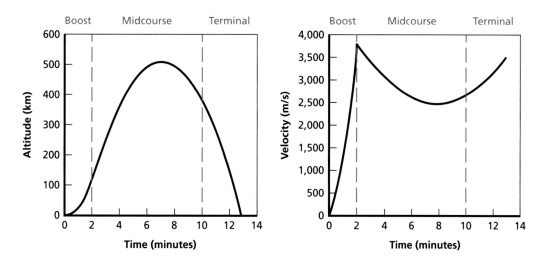

estimated trajectory forward to determine the missile's position at a given future time and begin their search for the missile from that point.

During a sensor blackout when no data are available to update the target state space, errors in the position estimate will build depending on its velocity, the duration of the blackout, and how far forward in time the trajectory is propagated. In the absence of other errors in modeling the missile's environment, the position error can be described thus:

$$\varepsilon = \int_{t=0}^{T} \left(V \cdot \ell \right) dt \, ,$$

where V is the missile velocity, ℓ is latency, and T is the trajectory propagation time window.

For the trajectory shown, Figure 7.6 shows this cumulative position error for each message (Figure 7.6a) and for a 2-minute propagation time (Figure 7.6b), as a function of latency (sensor blackout):

For example, if the latency introduced in the system is 1 microsecond (blue line), given the velocity profile of the target, the estimated accumulated position error is on the order of 1 m. Similarly, a latency of 1 ms (green line) will result in an accumulated position error of ~10 m. However, if the latency is as large as 1s (red line), the position error can be as large as 1 km. Depending on the sensor that is using these data (and being cued by the affected sensor) and its capabilities, this position error may mean

Figure 7.6
Error in Target Position for Different Amounts of Latency

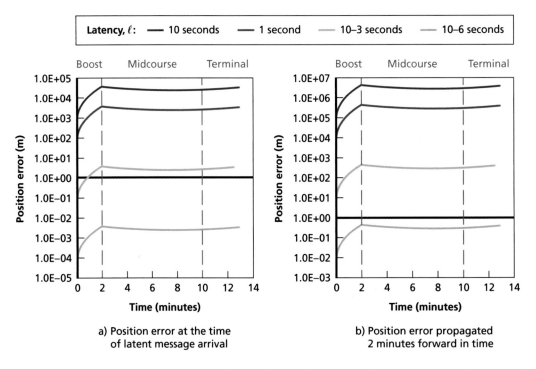

a) Position error at the time
of latent message arrival

b) Position error propagated
2 minutes forward in time

that the downstream sensor is unable to locate the target. For instance, a latency of 1 second (which results in approximately a 1 km position error) means that a space-based infrared sensor with an accuracy on the order of 500 m and a very narrow field of view, very close to the target, is not likely to locate the target based on a cue. However, for a typical kill vehicle sensor with a field of view of about 1 degree,[13] the target will still be in the field of view of the kill vehicle, given a 1 km uncertainty in target position, as long as the range to the target is greater than 58 km.

This high-level analysis shows that unless the latency introduced by the possible overhead required by a UCCL is on the order of 1 second, even in the scenario where sensor measurements or tracks are completely lost, BMD performance is not likely to be affected. The position error introduced by that latency is much less than the error that would affect the other systems that depend on the affected sensor. Several other factors can also contribute to this lack of impact: (1) during the midcourse phase, where most of the BMD systems are envisioned to operate, the missile is unpowered—

---

[13] National Research Council, 2012.

i.e., it is moving only under the forces of momentum and gravity, both of which are reasonably well modeled. Any lack of data during a fairly large time window will not affect the ability of the state estimators to estimate the missile's position and velocity with relatively low error; (2) because a total loss of messages is unlikely due to latency alone, measurements and tracks will still be available as input to the state estimators, which can use their time stamp to properly correlate them with existing tracks.

However, if the estimated latency introduced in the BMD system due to a UCCL is on the order of 1 millisecond to 1 second, and the kill vehicle is expected to have a sensor with a very narrow field of view and may receive a new track when fairly close to the target, then a more detailed analysis of the BMD system is required to properly model the state estimation, latency, and interactions among sensors and more accurately estimate the impact of latency on track quality and eventual P_{kill} of the BMD system.

Critical Step Analysis

Having identified interfaces in the current BMD architecture where use of a UCCL might be problematic or beneficial, we then turned our attention to examining a more generalized BMD architecture. The promise of a UCCL is that it will allow "any sensor, any shooter" joint operational capability. As seen in Figures 7.7–7.9, deployment of functions to physical items in the BMD architecture varies greatly. For instance, all radars can track the threat cloud, but only some can discriminate; some radars are designed to interface with a fire control system, but most are not. Table 7.2 maps top-level functions of a BMD system to physical system types; a large X indicates a primary function that all systems of this type have, and a small x indicates a function that some systems of this type currently have or are discussed as having in the future. This variety exists because system designers have had to respond over time to requirements for varying levels of autonomy and to different threats.

In a future conflict in which communication with a centralized C2 node is denied, we might imagine a BMD architecture in which any sensor with connectivity to the shooter (or interceptor once launched) is able to cooperatively negotiate the functions each has the resources and authority to perform and dynamically reconfigure the system topology to complete the mission. To realize this true "any sensor, any shooter" paradigm, a UCCL must be viable between each possible mapping of functions to systems. Therefore, we next examine how an abstraction of the interface between each of these functions might affect latency, network congestion, or computational burden.

Table 7.2
Mapping BMD Mission Functions to Physical Systems

BMD Function	Space Sensor	Land- and Sea-Based Sensor	Battle Management Center[a]	Shooter Fire Control System	Interceptor GNC[b]
Detect threat launch	X	x			
Cue sensors	x	x	X		
Track threat	X	X			x
Predict threat trajectory	x		X	x	X
Identify threat	x	x	X		
Discriminate		x	X	x	x
Acquire target		X			X
Approve fire[c]			X	X	
Guide interceptor				x	X

[a] Today's battle management centers are physically centralized. Geographic distribution or virtualization of the battle management function in the future may improve total system resilience.

[b] GNC = guidance, navigation, and control. While all kill vehicles have a GNC function, some are more autonomous than others. All of today's interceptors have an on-board seeker (generally infrared) as part of their GNC. Prior to seeker target acquisition, the on-board GNC relies on land- or sea-based sensor measurements as an input to the guidance loop.

[c] Nominally, a human on the loop will approve the decision to fire interceptor, but some systems (e.g., Aegis) have a fully autonomous mode that allows the fire control system to make these decisions. This autonomy may be critical if the system faces a salvo of incoming warheads.

The Mission Thread and Its Interfaces

We considered the potential utility and cost of implementing a UCCL at each of today's system-to-system interfaces as shown in Figures 7.7 to 7.9 for different phases of the BMD mission. This analysis provided three findings:

- The greatest potential system utility of implementing a UCCL is likely to be at the sensor interface and the fire control interface.
- The interface most likely to "break" if a UCCL is implemented is that between the fire control system and the kill vehicle.
- A UCCL that does not treat time as a critical architectural entity would break the BMD mission.

Figure 7.7
Communication Patterns for Threat Identification, Track Estimation, and Downstream Sensor Cueing in Current BMD Architecture

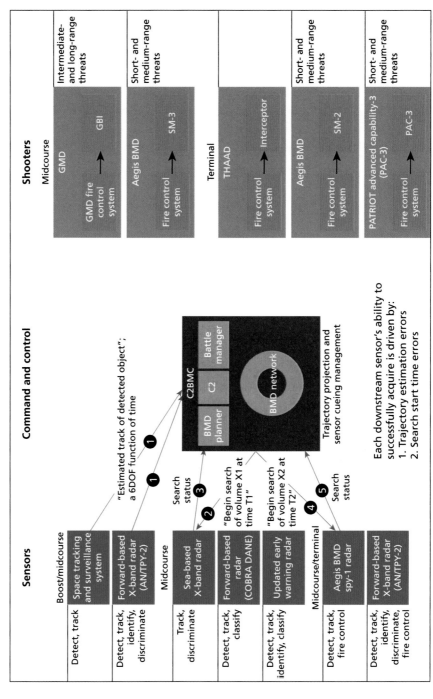

NOTE: GBI = Ground-Based Interceptor; GMD = Ground-Based Midcourse Defense.

The BMD System Interfaces

Figure 7.7 shows the primary communications needed for boost and early midcourse activities to identify and track the threat and to cue downstream sensors to join the sensor network. As noted earlier, obtaining additional sensor measurement from different geometries greatly enhances track accuracies. We hypothesize that if nontraditional sensors, such as those mounted on allied aircraft or ships, could be included in a future BMD sensor network on an ad hoc basis, the improvement in track accuracy and discrimination algorithm performance might be substantial. However, to achieve these benefits, the following must be true:

- Sensed measurements must use the same reference frame—including a time reference frame—and have a shared convention for reporting the time at which measurements were valid.
- The handover volume for the trajectory estimate must be less than the search volume of the cued sensors. Any differences in reference frame increase the uncertainty of the handover volume and the probability that a cued sensor will fail to acquire the threat cloud.[14] This is one more reason why the sensors need a shared convention regarding the definition of time.
- Measurement data must include measurement covariances unless the C2BMC has another source of knowledge from which to form a measurement covariance for each potential ad hoc sensor. We say more about the need for measurement covariances in the next section.

Figure 7.8 shows communication patterns needed for target discrimination. Each sensor report tracks for the threat cloud that includes the number of tracks and estimates for the track of each object. For sensors with on-board discrimination, an estimate of the lethality of each object in the threat cloud is also provided. The total set of information about the threat cloud is called a target object map (TOM). Midcourse discrimination may be the single most difficult task for a BMD system. Lethality becomes more observable after the objects reenter the atmosphere; however, intercepts at high altitude are preferred, to minimize collateral damage. Without talk-back and retarget capabilities onboard an interceptor, the decision between when to shoot and when to wait to collect more observations for discrimination will remain a very difficult one. Precision time—on the order of the clock drifts between GPS 1 pulse per second signals—will matter. In its 2012 report, the National Research Council noted:

---

[14] For the stressing case defined above and under worst-case search geometries, a 1 second time uncertainty regarding when to initiate search introduces an additional uncertainty of 5 km in the handover volume.

Figure 7.8
Communication Patterns for Target Discrimination in Current BMD Architecture

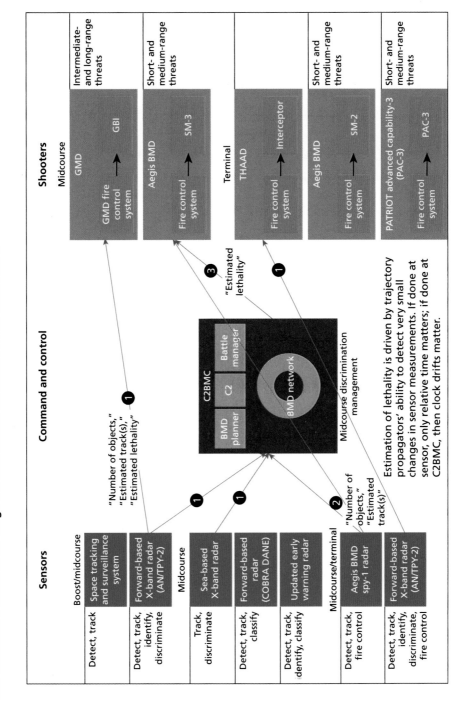

Data latency is a potential problem for the integrated battle command system (IBCS) that should not be ignored. [. . .] The evolutionary approach would employ much longer concurrent threat observation by both X-band radars and the interceptor KV's onboard sensor over the entire engagement. The importance of the synergy between these concurrent observations together with [Space Launch System] battle space in maximizing midcourse discrimination effectiveness cannot be overemphasized.[15]

As shown in Figures 7.8 and 7.9, the C2BMC currently has the capability to task Aegis, THAAD, Patriot, and GMD shooters through the use of a common interface specification called ATOMs.[16] The C2BMC-to-shooter interface is to the fire control system at the shooter, which in turn controls both the collocated radar (if there is one) and the interceptor. Using a UCCL between the fire control system and the interceptor would provide minimal utility because this is a tightly coupled interface, and each system partitions the interface differently to meet its unique mission.[17] Furthermore, if this interface uses the in-band RF signals of the radar as the link layer for fire control–to–interceptor communication, the National Research Council estimates that a system using in-flight target updates (IFTUs) would consume up to 65 percent of the radar's resources in the final seconds of flight.[18] Adding more overhead from a UCCL to this already resource-constrained communication might indeed "break" the system design.

While we do not recommend extending use of ATOMs for the interface between interceptor and fire control system, using it between the *terminal radar* and the fire control system has the potential to improve in-theater resilience. For instance, being able to use THAAD radar measurement to guide an Aegis missile without having to be connected through the C2BMC may provide greater resilience in a communications-denied environment.

---

[15] National Research Council, 2012.

[16] The C2BMC currently uses a common control language called ATOMs within the center. Sensor and shooter data are converted from and to its system-unique format when it enters and/or exits the center. New shooters are being asked to design to the ATOMs interface for the C2BMC-to–fire control system interface, eliminating the need for format conversions.

[17] For example, controlling a THAAD missile from a PATRIOT fire control system would require a significant rework of both of those systems. The effort needed might be better invested in making fire control systems more portable and survivable.

[18] National Research Council, 2012, p. 154.

Figure 7.9
Communication Patterns for Interceptor Engagement in Current BMD Architecture

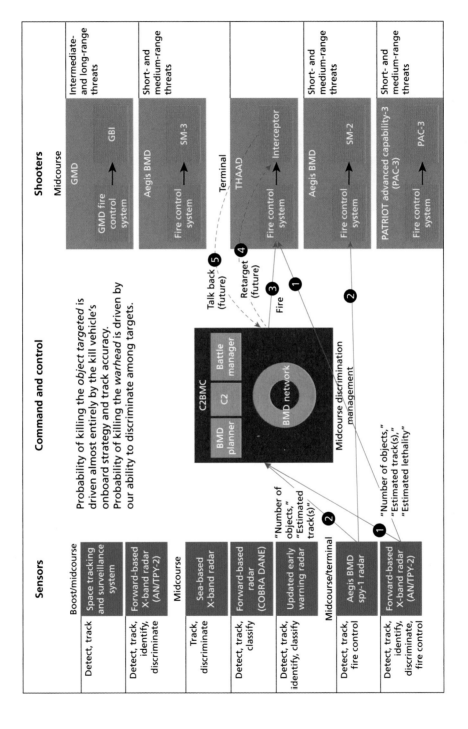

Sensor Interfaces

Today's space sensor network is largely composed of satellites in geosynchronous and highly elliptical orbits that individually send their information to the ground, where it is synthesized with other sensor data (space or terrestrial) for threat identification and cueing of downstream sensors. Some analysts envision a future space sensor network that utilizes resilient space or terrestrial networks to gather data from multiple payloads in multiple orbits and perhaps even from terrestrial or air-based sensors to perform data synthesis and cue downstream sensors (and perhaps even interceptors)[19] from space.[20] Architecturally, a fully capable sensor in such a system would produce and/or consume the following information:

1. threat detection event data
2. list of currently identified objects
3. track data for an object in the list
4. handover volume
5. search status.

The source and destination of this information, using the functional decomposition of Table 7.1, is shown in Figure 7.10. Some or all of these functions would be onboard the sensor itself or distributed across the physical architecture. We take a detailed look at three of the information items: threat detection event data, track data, and handover volume.

The ability of the SoS to correlate threat detection event data and track data will depend on a shared reference frame. For example, suppose sensor A reports an initial threat detected at position (latitude, longitude, altitude, time)$_A$, and sensor B soon thereafter reports an initial threat detected at (latitude, longitude, altitude, time)$_B$. Even if these state vectors are reported using identical reference frames (for example, Earth-centered inertial [ECI] for position and Universal Time Coordinated [UTC] for time), this information alone is not enough to determine whether this information represents two different missiles or two reports of the same missile. To make that determination, we also need an assessment of the state of health for the measure-

---

[19] Sandra Erwin, "Next Steps for the Pentagon's New Space Sensors for Missile Defense," *Space News*, January 21, 2019.

[20] While being able to defend from space has been a recurring theme throughout the history of missile defense, the two principal advantages of space are (a) its field of view and (b) its relative distance from the actual conflict. Given ongoing development and demonstration of antisatellite technologies and even fully operationalized weapons, these advantages may disappear. Therefore, in this section, we focus instead on the advantages of a distributed and fully meshed space-air-terrestrial architecture.

Figure 7.10
Key Sensor Information Flows for a Future BMD Architecture

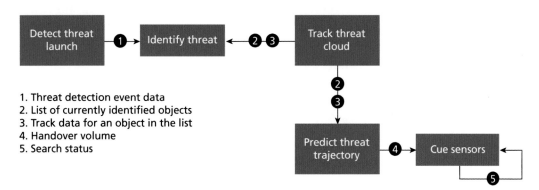

ment system and the sensitivity of the measurement to errors in each dimension of the reference frame: The latter is often conveyed as a covariance matrix.[21] We would also like to have the sensors tell us the velocity of the detected object to understand if back propagation of the detected event leads to the same launch location. Finally, we need each sensor to report a temporary *Threat ID* (to distinguish this track from other threats the sensor may be observing), which the SoS could replace later with an Assigned Threat ID. The total information needed for a *threat detection event* might be

> Sensor ID: integer
> Threat ID Type: temporary/assigned
> Threat ID: integer
> Threat State Vector: (position and velocity in three dimensions, time of validity)
> Measurement Integrity: healthy/suspect/unhealthy
> Measurement Covariance Matrix.

As an astute reader may have already observed, simply defining the UCCL information needed from a sensor for threat detection leads to a number of architectural design decisions that could affect the sensor's computing burden. If we go further and break the information into messages with syntax and format, we will also be making decisions that could influence congestion of the underlying meshed network. The following are the most significant decisions we made in defining the UCCL information:

---

[21]   A mobile sensor, such as that mounted on an aircraft or satellite, will need to know its own location with respect to the detected threat to calculate its measurement sensitivity. While today's C2BMD can receive that information from other sources to estimate the covariance, in a distributed system of the future, the covariance will need to be part of the detection event data.

- **Estimating the measurement covariance is the accountability of the sensor.** An alternative option might be to have the sensor communicate all the data that go into a measurement covariance calculation. This would include the sensor's own state vector, calibration curves of measurement accuracy versus relative geometry to the threat, and noise measurements in all degrees of freedom. While it may be a small burden today for the C2BMC to track this information for each sensor, in tomorrow's architecture it would become prohibitive to share that information across a sensor-shooter peer-to-peer distributed system.
- **In defining message syntax and format, the measurement covariance should be a separate message and transmitted as its inverse.** Designing the SoS in this manner would allow for distributed state estimation using these localized measurement covariances.[22] While distributed state estimation might be overkill for threat identification, using it for track estimation appears promising. For this reason, a UCCL might partition the sensor measurement covariance as its own message and specify that it be computed and transmitted as its inverse, suitable for use in a distributed iterate-collapse inversion (DICI) algorithm.[23]

Given the limited computing power of space platforms,[24] a UCCL interface for a space sensor might also be at a high risk of not having enough computational resources to execute the mission successfully and, therefore, would require a much more detailed study.

---

[22] For an example of how a distributed Kalman filter could be used to estimate state in a distributed sensor network, see Usmam Khan and Jose Moura, "Distributing the Kalman Filter for Large-Scale Systems," *IEEE Transactions on Signal Processing*, Vol. 56, No. 10, October 2008. This paper suggests that the inverse of the error covariances be distributed rather than the covariance itself. Note that transmitting full state estimates for a centralized state estimation has been shown in other distributed sensor applications to consume an unsupportable amount of network bandwidth.

[23] A Kalman filter is an implementation of Bayesian estimation using maximum a posteriori (MAP) techniques. While the Kalman filter has historically been used for state estimation in missiles and missile defense, other distributed estimators using MAP techniques also use the inverse measurement covariance in their implementation and may be lighter-weight. An example is given in Sun Yibling, Minue Fu, and Huanshui Zhang, "Performance Comparison of Distributed State Estimation Algorithms for Power Systems," *Journal of System Science Complexity*, Vol. 30, 2017.

[24] A general rule of thumb is that the best space-based computing architectures are two or three iterations of Moore's Law behind general usage computing architectures here on Earth.

Conclusions

BMD is a thread that takes minutes to execute from launch detection to final kill. Additional end-to-end latencies imposed by an inefficient interface design on the order of seconds are unlikely to influence the successful execution of the thread. But there are links within the thread with very constrained bandwidth, such as the SBIRS downlink, or the THAAD fire control link. These interfaces might require an optimized UCCL interface. For example, we identified a critical step in the thread that could affect P_{kill}, and that is the final sensor update to the interceptor. Latency in this step caused by a UCCL is *unlikely* to influence P_{kill} if elements of the BMD system have been calibrated for clock drift and include time of validity in sensor measurements, trajectory estimates, cueing, and targeting messages. However, it may influence warhead discrimination.

Additional considerations may influence performance. A UCCL may provide the opportunity to rapidly add sensors to the system. Such a diverse set of sensor geometries could improve trajectory estimates and warhead discrimination, improving P_{kill}. However, throughput and congestion of the underlying network caused by an inefficient UCCL could erase these gains. For example, some elements of BMD systems are resource-constrained. The following are examples:

- THAAD fire control link to interceptor is bandwidth-constrained, sharing radar resources; a UCCL may provide little benefit and substantial risk if used for this interface.
- State estimation to support trajectory estimation and warhead discrimination is likely to become processing-constrained as additional sensors are added; a UCCL that supports distributed state estimation could relieve this constraint, but the UCCL must be efficient.

If real-time composability of the sensor-to-shooter network is desired, network throughput and congestion are *likely* to be greatly affected by a UCCL's discovery protocol. The detailed implementation of a discovery protocol should be tailored to the needs of a particular mission to account for bandwidth-constrained elements of the system.

CHAPTER EIGHT

Conclusions

There are multiple examples of implementations of standards that require trade-offs between interface design and performance to achieve interoperability. These examples involve trade-offs in performance parameters such as delays, data rates, memory use, and data processing. In the case of DDS, we see how a different optimized adaptation of a standard can address some of these issues by trading encryption, security, and quality of service for speed; resiliency for memory and processing requirements; and network reconfigurability and flexibility for network bandwidth. These should be part of the technical considerations when designing any new standards.

Beyond the technical performance considerations, there are also important non-technical considerations. These primarily are related to the economics of standards that can enable network effects, better retention of human capital, reduced vendor lock, and cheaper training and retraining. While the technical performance of any new standard should be examined and analyzed to ensure its technical and operational viability, nontechnical considerations could lead to failed implementations of standards by failing to get broadly accepted in the market. This can be avoided by designing and implementing an effective and efficient standardization process, understanding stakeholders and their motivations, and understanding the market and the legal and regulatory context within which the standardization process will take place.

As indicated by results from past experiments and from our mission thread models, the implementation details of a standard interface could vary technical performance by orders of magnitude. The most pronounced effect is on system delays. Factors that influence delays are message size, data processing complexity, transfer time over a communication link, and how the messages flow from system to system. For cases in which the operational performance requirements are not very demanding, interface overhead will have only a minimal impact on operational performance. However, we showed that there are missions or steps within a mission where overhead may have a critical impact on mission performance. For instance, in the BMD thread, we saw a case of a single interceptor link that is severely bandwidth-limited, and any additional overhead may have significant impact on operational performance. In the APS thread, and for some higher-end threats, we may be forced to limit the amount of overhead we are willing to incur. For the EW thread, even modest assumptions may have a significant impact on mission performance. Therefore, any attempt at implementing a standard interface requires in-depth engineering analysis and careful design.

Systems implementation standards can mitigate some of the higher-risk systems and interfaces in some common ways:

- Focus interoperability on non–time-critical interfaces or interfaces with wide performance margins that take advantage of the benefits of more or better sensor-shooter pairing. The focus should be on areas estimated to lack strong dependencies between operational performance and message delays, and at the same time on those that have a better chance of providing higher operational benefits.
- For interfaces that have tighter delay and timing requirements, optimize the interface for compile time composability (packed versus unpacked). In other words, optimize the interface prior to the mission, not dynamically during the mission.
- Create a version of the standard that is applicable to performance-constrained systems.

When evaluating risks, consider mission-critical systems that have restrictive operational performance and timing requirements. Special attention should be given to these areas, and careful system performance analysis should be conducted to understand the operational risks associated with interface inefficiencies. Areas of concern should be

- systems with severely restricted bandwidth links
- systems with processors that have very little available processing power
- problems that require a lot of data operations per bit of data and possibly the processing of additional data mandated by the standard
- systems with submillisecond performance requirements.

While the modeling we performed in this analysis was intended to be a high-level investigation of the general trade-offs between interface inefficiencies and operational performance, our methodology could be applied in a more-detailed technical analysis of specific systems with the intent to model the real performance dependencies of the system by realistically modeling and evaluating its performance. The following are steps to evaluating the applicability of a universal standard interface to a particular system:

- Translate the mission-critical operational requirements into a system performance trade-space (e.g., delays).
- Quantify the overhead that a specific standard interface implementation would impose.
- Estimate the impact on operational performance and compare that with the operational requirement associated with the mission.
- Assess whether some of the impact could be mitigated by technical means—for example, by optimizing the particular interface or using an optimized version of the standard. If the issue is related to processing power, then adding more processors could be the solution. Where bandwidth is an issue, provide a higher bandwidth link.
- If the impact cannot be easily or cost-effectively mitigated by technical means, can it be mitigated by tactics, techniques, or procedures? For example, is it possible to maintain larger operational distance from the threat? To make changes to force structure to mitigate the threat? To modify the way systems are employed?
- Finally, if the operational performance limitation cannot be mitigated reasonably and cost-effectively, then the interface in question may not be a good candidate for standardization.

Publish and Subscribe Overview

Origins of the Publish-Subscribe (Pub-Sub) Model

In early network communications, computers communicating together would do so in one of two ways: either a broadcast to every point on the network or a direct communication to a specific other machine. As network and message complexity increased, the complexity of these interactions also became difficult to manage. Significant network capacity would be used up creating a link between two machines only to transfer a tiny status message. Those two machines would be busy during their interaction, causing other messages sent to either one to fail. Network programmers began to store messages and retry them when such failures occurred. Adding or removing machines in such a network was cumbersome because each new machine needed to be aware of all other machines, and removing a machine meant removing its status from each other machine. These behaviors created three problems: significant overhead in communications, common and unpredictable retry behavior, and complex and error-prone human management of the network.

To solve these problems and others, the *publish-subscribe* networking metaphor was adopted. Publish-subscribe is a communications metaphor commonly used in modern network communications protocols. It entails the creation of a storage space where network messages are held until a machine needing the information becomes available to read them. Instead of communicating directly with each other, machines communicated with a core server (or servers). These core servers' storage spaces can be grouped in various ways, but organizing them either by information source (e.g., for many machines to read the output of a weather sensor) or by topic (e.g., to allow any machine to report activity to a security logging service) is common. Publish-subscribe systems also often include a discovery subprotocol that allows a machine to be added to or removed from the network automatically. Publish-subscribe is not the only machine-to-machine communications model, but it is an extremely common one.

The Data Distribution Service

The *DDS protocol* has become one of the most commonly used publish-subscribe standards. It was developed between 2001 and 2004 by a pair of defense contractors and has spread to be used in a wide variety of capacities beyond its beginnings in aerospace and defense. Today, DDS is used in the Dutch railway management system, Volkswagen's smart cars, Siemens wind turbine fields in Iowa, Komatsu mining machines, General Electric medical scanners, and many other systems.

References

Augereau, Angelique, Shane Greenstein, and Marc Rysman, *Coordination Vs. Differentiation in a Standards War: 56k Modems*, NBER Working Paper 10334, Cambridge, Mass.: National Bureau of Economic Research, 2004. As of February 23, 2021:
http://www.nber.org/papers/w10334

Becker, Ofri, Joseph Ben Asher, and Ilya Ackerman, "A Method for System Interface Reduction Using N2 Charts," *Systems Engineering*, Vol. 3, 2000, pp. 27–37.

Bluetooth SIG, "Specifications," webpage, 2021. As of February 28, 2021:
https://www.bluetooth.com/specifications/

Cargill, Carl F., "Why Standardization Efforts Fail," *Journal of Electronic Publishing*, Vol. 14, No. 1, Standards, Summer 2011.

Church, Jeffrey, and Neil Gandal, "Network Effects, Software Provision, and Standardization," *Journal of Industrial Economics*, Vol. 40, No. 1, March 1992, pp. 85–103.

Committee on Armed Services, Tactical Air and Land Forces Subcommittee, *Combat Vehicle Active Protection Systems*, in U.S. House of Representatives, 109th Congress, second session, Washington, D.C.: U.S. Government Publications Office, September 1, 2006.

Deagle, homepage, undated-a. As of March 5, 2021:
http://www.deagel.com

Deagle, "KH-59," webpage, undated-b. As of March 5, 2021:
https://www.deagel.com/Offensive%20Weapons/Kh-59/a001024

Defense Industry Daily staff, "JLENS: Co-ordinating Cruise Missile Defense—and More," *Defense Industry Daily*, February 13, 2017. As of February 23, 2021:
https://www.defenseindustrydaily.com/jlens-coordinating-cruise-missile-defense-and-more-02921/

De Weck, Olivier L., "Fundamentals of Systems Engineering," lecture notes slide deck, Massachusetts Institute of Technology Lincoln Labs, Fall 2015. As of February 23, 2021:
https://ocw.mit.edu/courses/aeronautics-and-astronautics/16-842-fundamentals-of-systems-engineering-fall-2015/lecture-notes/MIT16_842F15_Ses_8_Sys_Int.pdf

Eleazer, Wayne, "Launch Failures: The 'Oops!' Factor," *Space Review*, January 31, 2011. As of February 23, 2021:
https://www.thespacereview.com/article/1768/2

eProsima, "Micro XRCE-DDS," webpage, 2019. As of February 23, 2021:
https://www.eprosima.com/index.php/products

Erwin, Sandra, "Next Steps for the Pentagon's New Space Sensors for Missile Defense," *Space News*, January 21, 2019.

Falcone, Stephen J., "Modular Open Systems Architecture/Approaches," briefing slides, U.S. Air Force Life Cycle Maintenance Center Battle Management Directorate, February 6, 2018.

Feickert, Andrew, "Army and Marine Corps Active Protection System (APS) Efforts," Washington D.C., August 30, 2016. As of February 23, 2021:
https://digital.library.unt.edu/ark:/67531/metadc958725/

"FIM-92 Stinger MANPADS Man-Portable Surface-to-Air Missile System," fact sheet, Army Recognition.com, November 13, 2020. As of March 4, 2021:
https://www.armyrecognition.com/united_states_american_missile_system_vehicle_uk/stinger_
fim-92_fim-92a_man_portable_air_defense_missile_system_manpads_technical_data_sheet_
picture.html

General Dynamics Mission Systems, "TACLANE-Micro (KG-175D) Encryptor," fact sheet, 2020. As of February 23, 2021:
https://gdmissionsystems.com/products/encryption/taclane-network-encryption/
taclane-micro-kg-175d-encryptor

GlobalSecurity, "Raduga Kh-59 (AS-13 Kingbolt) and Kh-59M (AS-18 Kazoo)," webpage, updated October 18, 2015. As of March 5, 2021:
https://www.globalsecurity.org/military/world/russia/as-13.htm

Gompert, David C., and Jeffrey A. Isaacson, *Planning a Ballistic Missile Defense System of Systems: An Adaptive Strategy*, Santa Monica, Calif.: RAND Corporation, IP-181, 1999. As of February 23, 2021:
https://www.rand.org/pubs/issue_papers/IP181.html

"Hardkill APS Overview," *Below the Turret Ring* blog, January 7, 2017. As of February 23, 2021:
https://below-the-turret-ring.blogspot.com/2017/01/hardkill-aps-overview.html

Haug, D., and H. J. Wagner, "Active Hard Kill Protection Systems—Analysis and Evaluation of Current System Concepts," *Strategie & Technik*, Autumn 2009.

Information Sciences Institute, University of Southern California, *Internet Protocol: DARPA Internet Program Protocol Specification*, Request for Comments 791, Arlington, Va.: Defense Advanced Research Projects Agency, Information Processing Techniques Office, September 1981. As of February 25, 2021:
https://tools.ietf.org/html/rfc791

"Instructions per Second," *Wikipedia* entry, webpage, updated February 22, 2021. As of March 5, 2021:
https://en.wikipedia.org/wiki/Instructions_per_second

"Interface Description Language," *Wikipedia* entry, webpage, updated January 4, 2021. As of February 25, 2021:
https://en.wikipedia.org/wiki/Interface_description_language

International Organization for Standardization and the International Electrotechnical Commission, "Open Systems Interconnection—Basic Reference Model: The Basic Model–Part 1," ISO/IEC 7498-1:1994, webpage, 1994. As of February 23, 2021:
https://www.iso.org/obp/ui/#iso:std:iso-iec:7498:-1:ed-1:v2:en

———, "Systems and Software Quality Requirements and Evaluation (SQuaRE)—System and Software Quality Models," ISO/IEC 25010:2011, webpage, 2011. As of February 23, 2021:
https://www.iso.org/obp/ui/#iso:std:iso-iec:25010:ed-1:v1:en

Isbell, Douglas, Mary Hardin, and Joan Underwood, "Mars Climate Orbiter Team Finds Likely Cause of Loss," press release, National Aeronautics and Space Administration, 1999. As of February 23, 2021:
https://mars.nasa.gov/msp98/news/mco990930.html

ISO/IEC—*See* International Organization for Standardization and the International Electrotechnical Commission.

Katz, Michael, and Carl Shapiro, "Systems Competition and Network Effects," *Journal of Economic Perspectives*, Vol. 8, No. 2, Spring 1994, pp. 93–115.

Keller, John, "Harris to Provide Military Special Operations Radios for Sensitive and Covert Missions," *Military and Aerospace Electronics*, January 22, 2014.

Kempinski, B., and C. Murphy, "Technical Challenges of the U.S. Army's Ground Combat Vehicle Program," in Isak Lundgren, ed., *The Army's Ground Combat Vehicle (GCV) Program*, Hauppauge, N.Y.: Nova Science Publishers, 2013, pp. 55–122.

Khan, Usmam, and Jose Moura, "Distributing the Kalman Filter for Large-Scale Systems," *IEEE Transactions on Signal Processing*, Vol. 56, No. 10, October 2008.

Kingsley, Simon, and Shaun Quegan, *Understanding Radar Systems*, Vol. 2, Chennai, India: SciTech Publishing, 1999.

Leonardo DRS Company, "IM-SHORAD," webpage, 2020. As of February 23, 2021:
https://www.leonardodrs.com/products-and-services/im-shorad/

Libicki, Martin C., James Schneider, David R. Frelinger, and Anna Slomovic, *Scaffolding the New Web: Standards and Standards Policy for the Digital Economy*, Santa Monica, Calif.: RAND Corporation, MR-1215-OSTP, 2000. As of February 23, 2021:
https://www.rand.org/pubs/monograph_reports/MR1215.html

Mackie-Mason, Jeffrey, and Janet Netz, "Manipulating Interface Standards as an Anticompetitive Strategy," in S. Greenstein and V. Stango, eds., *Standards and Public Policy*, Cambridge, UK: Cambridge University Press, July 2007.

Matutes, Carmen, and Pierre Regibeau, "Mix and Match: Product Compatibility Without Network Externalities," *RAND Journal of Economics*, Summer 1988.

Missile Defense Project, "Hong Niao Series (HN-1/-2/-3)," *Missile Threat*, database, Center for Strategic and International Studies, August 12, 2016, last modified November 26, 2019. As of May 4, 2021:
https://missilethreat.csis.org/missile/hong-niao/

Mohammad, Khader, Temesghen Tekeste, Baker Mohammad, Hani Saleh, and Mahran Qurran, "Embedded Memory Options for Ultra-Low Power IoT Devices," *Microelectronics Journal*, Vol. 93, November 2019.

National Information Exchange Model, "Movement," database, undated. As of February 25, 2021: https://beta.movement.niem.gov/#/results?q=*&selectedFacets=domain:%22Military%20 Operations%22&page=1

National Research Council, *Realizing the Information Future: The Internet and Beyond*, Washington, D.C.: The National Academies Press, 1994. As of February 25, 2021: https://www.nap.edu/catalog/4755/realizing-the-information-future-the-internet-and-beyond

———, *Making Sense of Ballistic Missile Defense: An Assessment of Concepts and Systems for U.S. Boost-Phase Missile Defense in Comparison to Other Alternatives*, Washington, D.C.: National Academies Press, 2012.

NIEM—*See* National Information Exchange Model.

Object Management Group, "Data Distribution Service Specification Version 1.4," spec sheet, March 2015.

Pan, Runyu, and Gabriel Parmer, "MxU: Towards Predictable, Flexible, and Efficient Memory Access Control for the Secure IoT," *ACM Embedded Computing System*, Vol. 18, No. 5s, Article 103, October 2019.

Poisel, Richard, *Modern Communications Jamming: Principles and Techniques*, Washington, D.C.: Artech House, 2011.

Popenker, Maxim, "Modern Firearms," webpage, undated. As of March 5, 2021: https://modernfirearms.net/en/

PubNub, "What Is Publish-Subscribe (Pub/Sub)?" *Realtime Technology Glossary*, website, undated. As of February 25, 2021: https://www.pubnub.com/learn/glossary/what-is-publish-subscribe/

Ramjug, Peter, "Raytheon's Quick Kill Active Protection System Defeats One of the Most Lethal Armor-Piercing Rocket Propelled Grenades," press release, Raytheon Co., January 9, 2013. As of March 7, 2021: https://raytheon.mediaroom.com/index.php?s=43&item=2251

Resch, Cheryl, "Exo-atmospheric Discrimination of Thrust Termination Debris and Missile Segments," *Johns Hopkins APL Technical Digest*, Vol. 19, No. 3, 1998.

Rohatgi, Ankit, WebPlotDigitizer, Version 4.4, web-based tool, November 28, 2020. As of February 28, 2021: https://automeris.io/WebPlotDigitizer/

RSA Laboratories, "FAQ: What Is a Block Cipher?" webpage, 1998. As of February 23, 2021: http://security.nknu.edu.tw/crypto/faq/html/2-1-4.html

Rudd, Kevin, "CONverged Collaborative Elements for RF Task Operations (CONCERTO)," webpage, Defense Advanced Research Projects Agency, undated. As of December 16, 2021: https://www.darpa.mil/program/converged-collaborative-elements-for-rf-task-operations

Rutledge, Edward, et al., *UCCL Performance Study Report*, Cambridge, Mass.: Massachusetts Institute of Technology Lincoln Labs, forthcoming.

Rysman, Marc, "Adoption Delay in a Standards War," thesis, Boston University, October 2003.

Simcoe, Timothy S., Stuart J. H. Graham, and Maryann Feldman, "Competing on Standards? Entrepreneurship, Intellectual Property, and the Platform Paradox," NBER Working Paper 13632, Cambridge, Mass.: National Bureau of Economic Research, November 2007. As of February 23, 2021:
https://www.nber.org/papers/w13632.html

Steeb, R., *Issues for Ground Vehicle Active Protection Systems for the Next Decade*, draft report U.S. Army Fort Benning Maneuver Center of Excellence, 2017.

Sun, Yibling, Minue Fu, and Huanshui Zhang, "Performance Comparison of Distributed State Estimation Algorithms for Power Systems," *Journal of System Science Complexity*, Vol. 30, 2017, pp. 595–615.

SyntheSys Defence, "Variable Message Format (VMF)," fact sheet, undated. As of December 28, 2021:
http://www.synthesys.co.uk/brochures_new/VMF%20Information%20Sheet.pdf

Tucker, R. S., and K. Hinton, "Energy Consumption and Energy Density in Optical and Electronic Signal Processing," *IEEE Photonics*, Vol. 3, 2011, pp. 821–833.

Turner, Mark, and Ken Dingman, "Developing SCA Based Wideband Networking Waveforms," presentation slides, Harris Corporation, 2011. As of February 23, 2021:
https://www.wirelessinnovation.org/assets/Proceedings/2011/2011-7d-dingman-presentation.pdf

U.S. Air Force, "Universal C2 Interface, Part of the Open Mission Systems," briefing, 2017a.

———, "Open Mission Systems (OMS)," briefing, September 27, 2017b.

U.S. Marine Corps, "Ground Based Air Defense," webpage, undated. As of December 17, 2021:
https://www.peols.marines.mil/Programs/Ground-Based-Air-Defense/

Upton, Lee O., and Lewis A. Thurman, "Radars for the Detection and Tracking of Cruise Missiles," *Lincoln Laboratory Journal*, Vol. 12, No. 2, 2000.

"Use of Open Mission Systems/Universal Command and Control Interface," Department of the Air Force memorandum for Air Force Program Executive Officers, October 9, 2018.

Viasat, "MIDS-LVT Terminals," webpage, 2021. As of March 5, 2021:
https://www.viasat.com/products/terminals-and-radios/mids-lvt/

Weiss, Martin B. H., and Marvin Sirbu, "Technological Choice in Voluntary Standards Committees: An Empirical Analysis," in *Economics of Innovation and New Technology*, Vol. 1–2, 1990, pp. 111–133.

Werrell, Kenneth, "Hitting a Bullet with a Bullet: A History of Ballistic Missile Defense," Airpower Research Institute Research Paper 2000-02, Air University College of Aerospace Doctrine, Research and Education, Maxwell Air Force Base, Alabama, 2000.

Wong, L., "Systems Engineering Approach to Ground Combat Vehicle Survivability in Urban Operations," thesis, Naval Postgraduate School, Monterey, Calif., 2016.

Xiong, Ming, Jeff Parsons, James Edmondson, Hieu Nguyen, and Douglas C. Schmidt, *Evaluating the Performance of Publish/Subscribe Platforms for Information Management in Distributed Real-Time and Embedded Systems*, Nashville, Tenn.: Vanderbilt University, 2011. As of February 23, 2021: https://www.dds-foundation.org/sites/default/files/Evaluating_Performance_Publish_Subscribe_Platforms.pdf